Springer Monographs in Mathematics

László Székelyhidi

Discrete Spectral Synthesis and Its Applications

 Springer

László Székelyhidi
University of Debrecen
Institute of Mathematics
H-4010 Debrecen, Hungary
email: szekelyhidi@math.klte.hu

A C.I.P. Catalogue record for this book is available from the Library of Congress.

ISBN-10 94-007-8723-5
ISBN-13 978-94-007-8723-0
ISBN-10 1-4020-4637-5 (eBook)
ISBN-13 978-1-4020-4637-7 (eBook)

Published by Springer,
P.O. Box 17, 3300 AA Dordrecht, The Netherlands.

www.springer.com

Printed on acid-free paper

To my sons and my wife

Contents

Preface

Spectral analysis and spectral synthesis deal with the description of translation invariant function spaces over locally compact Abelian groups. Translation invariant function spaces appear in several different contexts: linear ordinary and partial difference and differential equations with constant coefficients, theory of group representations, classical theory of functional equations, etc. A fundamental problem is to discover the structure of such spaces of functions, or more exactly, to find an appropriate class of basic functions, the building blocks, which serve as "typical elements" of the space, a kind of basis. It turns out that these building blocks are the so-called exponential monomials. A famous pioneering result of L. Schwartz [55] typifies this construction. One considers the space of all complex-valued continuous functions on the real line which is a locally convex topological linear space with respect to the pointwise linear operations (addition, multiplication with scalars) and to the topology of uniform convergence on compact sets. Suppose that a closed linear subspace in this space is given and that it is translation invariant. This subspace may or may not contain any basic function of the above mentioned form, that is, an exponential monomial, which is the product of a power function and an exponential function. It is not even clear if any exponential function is included in such subspaces. If it is so, then we say that *spectral analysis holds for the subspace* in question. It is not very difficult to show that the appearance of an exponential monomial in a subspace of this type implies that the exponential function occurring in this exponential monomial must belong to the same subspace, too. The complex number characterizing this exponential function can be considered as a kind of spectral value. A fundamental consequence of Schwartz's result is that spectral values do exist, meaning that any closed translation invariant linear subspace of the space mentioned above contains an exponential. The complete description of the subspace, however, means that all the exponential monomials corresponding to the spectral exponentials and their multiplicities characterize the subspace: their linear hull is dense in the subspace. If this is so, then we say that *spectral synthesis holds for the subspace*. In fact this is Schwartz's result: any closed translation invariant linear space of continuous functions on the reals is synthesizable from its exponential monomials.

The construction presented in this example can be generalized. Instead of the topological group of the reals, one considers the set of all continuous complex-valued functions on a locally compact Abelian group equipped with the pointwise linear operations and with the topology of uniform convergence on compact sets. In order to set up the problem of spectral analysis and spectral synthesis in this context, one has to define exponential functions and exponential monomials on commutative topological groups. Continuous homomorphisms of these groups into the additive topological group of complex numbers and into the multiplicative topological group of nonzero complex numbers are called additive and exponential functions, respectively. A polynomial is a linear combination of products of additive functions and constants, and an exponential monomial is a product of a polynomial and an exponential function. Now the problems of spectral analysis and spectral synthesis can be formulated: is it true that any nonzero closed translation invariant linear subspace of the space mentioned above contains an exponential function (spectral analysis), and is it true that in any subspace of this kind the linear hull of all exponential monomials is dense (spectral synthesis)? Here is the point where one can realize the special importance of some basic functions: exponential functions, additive functions, polynomials and exponential polynomials. As these function classes are fundamental also in the theory of functional equations, no wonder that there are several connections between functional equations and spectral analysis and synthesis. To give an insight into the nature of these connections is one of the main purposes of this volume.

Returning to the above situation it is easy to see that we can go one step further: instead of the space of continuous functions with the given topology we can start with other important function spaces that are translation invariant. For instance, the space of integrable functions is the natural medium of the Wiener Tauberian theory: different versions of the Wiener Tauberian Theorem can be stated as spectral analysis theorems. Nevertheless, in case of integrability all polynomials reduce to constants and in most cases exponentials are only the characters, hence we have a rather special situation.

An interesting particular case is presented by discrete Abelian groups. Here the problem seems to be purely algebraic: all complex functions are continuous and convergence is meant in the pointwise sense. The archetype is the additive group of integers: in this case the closed translation invariant function spaces can be characterized by systems of homogeneous linear difference equations with constant coefficients. It is known that these function spaces are spanned by exponential monomials corresponding to the characteristic values of the equation, together with their multiplicities. In this sense the classical theory of homogeneous linear difference equations with constant coefficients can be considered as spectral analysis and spectral synthesis on the additive group of integers.

The next simplest instance is the case of systems of homogeneous linear difference equations with constant coefficients in several variables, or in other words, spectral analysis and spectral synthesis on free Abelian groups with a finite number of generators. As in this case a structure theorem is available, namely, any group of this type is a direct product of finitely many copies of the additive group of integers, it is not very surprising to have the corresponding — nontrivial — result by M. Lefranc [35] : on finitely generated free Abelian groups both spectral analysis and spectral synthesis hold for any nonzero closed translation invariant subspace.

Based on the above mentioned results the natural question arises : do these concepts hold on arbitrary discrete Abelian groups? To make a strict distinction among different types of spectral analysis and spectral synthesis problems sometimes we shall use the term "discrete spectral analysis" and "discrete spectral synthesis". By this we want to make it clear that we are talking about spectral analysis and spectral synthesis problems on discrete Abelian groups, where the underlying function space is the space of all complex-valued functions equipped with the pointwise linear operations and with the topology of pointwise convergence. In his 1965 paper [16] R. J. Elliot presented a theorem on discrete spectral synthesis for arbitrary Abelian groups. In 1986, however, Z. Gajda [19] observed that the proof of Elliot's theorem had some gaps. Despite different efforts made by Gajda, the present author and others, these gaps have been impossible to fill. Until very recently even the question about discrete spectral analysis has remained open. However, in the last two decades new efforts resulted in a couple of interesting new methods and theorems on discrete spectral analysis and spectral synthesis. The study of further various applications in the theory of general functional equations and in other branches of mathematics has proved to be very fruitful. In this volume we give a survey on the theory of discrete spectral analysis and spectral synthesis and exhibit the most recent results together with various applications.

In the monograph [66] the present author attempted to exhibit wide-range applications of discrete spectral analysis and spectral synthesis in the theory of convolution type functional equations. Since then further applications have been presented on different areas, such as digital filtering (see [52], [53], [54]), eigenfunctions of difference operators (see [67]) and polynomial hypergroups (see [72], [77]). Applying discrete spectral synthesis to convolution type functional equations, a long-standing conjecture of H. Haruki and D. Z. Djokovič on mean value type functional equations has been proved (see [66], [71]).

In [70] we offered a possible way to prove discrete spectral synthesis for free Abelian groups by presenting an equivalent reformulation of the problem, but a proof for this equivalent version was not available. In [69] discrete spectral analysis for Abelian torsion groups was proved. In 2001 G. Székelyhidi (see [61]) presented a different approach to the result of Lefranc, and his method strongly supported the conjecture that discrete spectral analysis could hold on countably generated free Abelian groups, but could not hold for free Abelian groups having no denumerable generating set. At the 41st International Symposium on Functional Equations in Noszvaj, Hungary, in 2003 we presented a counterexample for discrete spectral synthesis on some special Abelian groups, disproving the result of Elliot in [16]. Recently in [32] M. Laczkovich and G. Székelyhidi gave the complete characterization of Abelian groups having discrete spectral analysis: they are exactly the ones having torsion free rank less than the continuum. On the other hand, a more detailed study of the above mentioned counterexample led to the main result in [73] which says that discrete spectral synthesis fails to hold on any Abelian group having infinite torsion free rank. In the light of Lefranc's result one may think that discrete spectral synthesis holds exactly on finitely generated Abelian groups. However, this is not the case: it has been proved in [3] that discrete spectral synthesis holds on any commutative torsion group. Hence the following conjecture seems to be reasonable to formulate as it has been done in [73]: discrete spectral synthesis holds on an Abelian group if and only if its torsion free rank is finite. Although this problem still remains unsolved, in [75] an equivalent formulation of this conjecture is presented in terms of polynomial functions: discrete spectral synthesis holds on an Abelian group if and only if any complex generalized polynomial on the group is a polynomial. This formulation of the basic problem of discrete spectral synthesis underlines again the strong connection with the theory of functional equations.

The basis of this volume is the monograph [66], however the present work is a highly updated version of that book. Our presentation here is partly based on a seminar given by the author during the 2002 Spring Semester at the Department of Mathematics, University of Louisville, Kentucky. The author is highly indebted to the colleagues who participated in the seminar's work for their valuable and encouraging help.

Besides this preface the present volume consists of a table of contents, an introductory first chapter and seven additional chapters completed with a reference list and an index.

The first chapter is the Introduction, where we exhibit the classical background of spectral analysis and spectral synthesis in functional analysis. Spectral problems arise in different areas of functional analysis and of representation theory; in some cases these problems can be converted into pure algebraic problems by using Gelfand–like transformations. In this introductory chapter we go through the most important classical problems of Gelfand theory, mainly by mentioning the results without proofs; however, the interested reader is provided with the necessary references.

In Chapter 2 the basic problems of spectral analysis and spectral synthesis are presented in different situations together with possible solutions, where the results are mainly stated without proofs. The most general setup shows the connections to the classical Wiener Tauberian theory, which is closely related to the ideal theory of L^1-spaces. Here we also point out the relationship with abstract harmonic analysis over locally compact Abelian groups. The corresponding results in L^∞-theory relate to the classical results of A. Beurling, P. Malliavin and to the Primary Ideal Theorem as it is presented in Section 2.3. Finally, the concluding section is devoted to continuous spectral synthesis, including results of B. Malgrange and L. Ehrenpreis. This chapter provides the motivation for investigating discrete spectral analysis and spectral synthesis by presenting its classical roots in different settings.

In Chapter 3 different problems concerning spectral analysis and spectral synthesis on discrete Abelian groups are studied. This is the heart of the book in the sense that the most recent results on the subject are collected here. Section 3.1 is devoted to spectral analysis on torsion groups. The result in [69], which shows that discrete spectral analysis holds on any commutative torsion group, represents the first general discrete spectral analysis result for Abelian groups having no finite system of generators. This theorem can also be interpreted as a general Wiener Tauberian theorem in an abstract situation. Section 3.2 presents one of the newest results in the theory of discrete spectral analysis. The characterization theorem of those Abelian groups on which discrete spectral analysis holds, given in [32], explains the extreme speciality of torsion groups from the point of view of discrete spectral analysis: they have torsion free rank zero. In Section 3.3 another special case is studied: the one represented by the additive semigroup of nonnegative integers and direct powers of it. It turns out that Lefranc's results can be generalized to this case. This leads to a nice characterization of polynomial ideals in several variables in Section 3.4 using systems of partial differential operators. The results in [76] are related to the Ehrenpreis–Palamodov Theorem and give a simple possibility to construct a set of Noetherian operators for a polynomial ideal in several variables. A crucial point of this chapter is Section 3.5: here we show that discrete spectral synthesis does not hold for a general class of discrete Abelian groups. The above mentioned counterexample we present here indicates a close relation between Abelian groups possessing spectral synthesis, and those having a "finite-like" generating set, or some other finiteness property. We can formulate the corresponding result in a way which clears up the relation between discrete spectral analysis and spectral synthesis in the light of the result of [32]: if discrete spectral synthesis holds on an Abelian group, then its torsion free rank is finite. One obvious consequence is that there are Abelian groups on which discrete spectral analysis holds, but discrete spectral synthesis fails to hold. At this point one may think that Lefranc's result characterizes Abelian groups with spectral synthesis: they might be exactly the finitely generated ones. However, Section 3.6 disproves this conjecture: by the results of [3], discrete spectral synthesis holds on any commutative torsion group, hence there are Abelian groups with discrete spectral synthesis and without a finite generating system. The only reasonable conjecture is the one formulated in Section 3.5: discrete spectral synthesis holds on an Abelian group if and only if its torsion free rank is finite. In Section 3.7 an equivalent reformulation of this conjecture is given which enlightens the relation with polynomial functions: it turns out that an Abelian group has finite torsion free rank if and only if any complex generalized polynomial on the group is actually a polynomial. Hence the failure of discrete spectral synthesis may be due to the existence of "pathological" polynomial functions on the group. However, this problem remains open.

Chapter 4 is devoted to applications of spectral synthesis for different types of functional equations. In Section 4.1 it is shown how to handle convolution type functional equations in the lack and in the presence of discrete spectral synthesis. In particular, equivalence and implication properties between systems of convolution type functional equations are studied here by presenting the results of [68]. In Section 4.2 we solve a long-standing conjecture of Haruki and Djoković for mean value type functional equations. The solution depends completely on discrete spectral synthesis. The corresponding result has been published in [71]. In Section 4.3 another special application of spectral synthesis is presented concerning a functional equation in digital filtering. The corresponding results have been published in [52], [53] and [54], representing how to solve particular problems in the presence of spectral synthesis.

Chapter 5 is devoted to the theory of mean periodic functions on Abelian groups. We start the study with the classical case of the real line equipped with the Euclidean topology. The famous result of Schwartz about continuous spectral synthesis on the reals, as it is presented in [55], is the basis of these investigations. We introduce a Fourier–like transformation for mean periodic functions on the real line in Section 5.1 and — using the same ideas — for exponential polynomials on arbitrary Abelian groups in Section 5.2, and we exhibit the most important properties of this transformation from the point of view of functional equations. Some special applications for ordinary and partial differential equations are also presented in Section 5.3. The first applications of this type appeared in [63], [64] and [65].

Applications of spectral synthesis to finitely generated free Abelian groups are studied in Chapter 6. The main emphasis is on difference equations. The classical theory of linear homogeneous difference equations with constant coefficients appears as a special case of spectral synthesis on the additive group of integers, and extensions of this theory to similar problems in several variables are treated. We offer a general method for finding the spectral set of linear homogeneous difference equations with constant coefficients in several variables which can be utilized to find the general solutions. In Section 6.1 the fundamentals of this method are presented in terms of spectral synthesis on finitely generated free Abelian groups, and in Section 6.2 applications to some special difference and recurrence equations are given as an illustration of the method. The corresponding results have been published in [74].

The last two chapters are devoted to some new results on spectral analysis and spectral synthesis over polynomial hypergroups. The author feels that these results represent a field which might have some interest in the future from the point of view of both spectral synthesis and of hypergroups. In Chapter 7 the corresponding results are included on polynomial hypergroups in a single variable and in Chapter 8 these results are extended to the multivariate case. The one-variable method has been published in [72] and the several-variable results are under publication in [77].

The purpose of this volume is to present wide-range applications of spectral analysis and spectral synthesis in different fields of mathematics, but the main emphasis is on applications in functional equations. The interested reader may receive an insight about relevant research in the theory of functional equations from several books and research papers published recently. The volumes [28] and [29] and the research papers [27] and [48] present the newest methods of stability theory of functional equations, while [47] and [9] offer a more detailed collection of recent results on functional equations in several variables.

The research and writing of this work were carried out while I was a visiting professor at Mississippi State University in Mississippi State during the 2002–03 academic year. I am highly indebted to Bruce Ebanks, Chair of the Department of Mathematics and Statistics at Mississippi State University that time, who offered ideal circumstances to carry out this work as a chief, as a mathematician and as a friend.

1

Introduction

The basic tools for the investigation of different algebraic and analytical structures are representation and duality. "Representation" means that we establish a correspondence between our abstract structure and a similar, more particular one. Usually this more particular structure, the "representing" structure is formed by functions, defined on a set which is the so-called "dual" object. In order to get a "faithful" representation, it seems to be reasonable that the correspondence in question is one-to-one. Another reasonable requirement is that if the same procedure is applied to the dual object, then its dual can be identified with the original structure. In order to do that, the dual object should have an "internal" characterization. Finally, a characterization of the "representing" structure is also desirable: which functions on the dual object belong to the "representing" structure?

The method of representation and duality appears in several different fields of algebra, analysis, etc. For instance, linear spaces can be represented as linear spaces of linear functionals, topological spaces can be represented as topological spaces of continuous functions, topological groups can be represented as topological groups of special homomorphisms, and so on. However, in all these cases one can assure the faithfulness via different assumptions only. In the case of linear spaces the injectivity of the representing mapping holds only if the linear functionals of the original linear space form a separating family, which leads to Hahn–Banach type theorems. In the case of topological spaces the same requirement leads to conditions similar to those in Uryshon's Lemma. In the theory of algebras the corresponding representation process can be described by the Gelfand transformation.

Let A be a complex algebra and let H denote a set of algebra homomorphisms of A onto $\mathbb{C}$, the algebra of complex numbers. Such homomorphisms are called *multiplicative linear functionals* . We remark that the assumption on the surjectivity of a complex algebra homomorphism is obviously equiva-

1

lent to it being nonidentically zero. Evidently, H is a subset of the algebraic dual of A, however, in general, H has no natural algebraic structure.

The Gelfand transformation on A is defined by

$$\hat{x}(h) = h(x)$$

for all x in A and h in H. Then $\hat{x} : H \to \mathbb{C}$ is a complex-valued function on H and obviously $x \mapsto \hat{x}$ is an algebra homomorphism of A onto the function algebra of all functions of the form $\hat{x}$, which we denote by $\widehat{A}$. The function $\hat{x}$ is called the *Gelfand transform* of x, and the mapping $x \mapsto \hat{x}$ is called the *Gelfand transformation*.

The representation of A by the function algebra $\widehat{A}$ is "faithful" in the above sense if and only if the Gelfand transformation is injective. It is easy to see that this happens if and only if H is a *separating family* for A, that is, for all $x \neq y$ there exists an h in H with $h(x) \neq h(y)$. As H consists of homomorphisms, it is enough to have an h in H for any nonzero x in A with $h(x) \neq 0$. Alternatively, we express this property by saying that A has *sufficiently many* multiplicative linear functionals. This means that the nonzero elements of A in the intersection of the kernels of all elements of H will violate injectivity. As obviously any element of the form $xy - yx$ belongs to this intersection, it is necessary for the injectivity of the Gelfand transform that A be *commutative*. From now on we shall suppose this, which seems to be a quite natural requirement if we want to represent A by an algebra of complex-valued functions, as such function algebras are commutative. However, commutativity of A is not sufficient for the Gelfand transform to be one-to-one.

The next step is to try to identify H. It is obvious that for any h in H the kernel of h is a *maximal ideal* in A. Maximality follows from the *Homomorphism Theorem*: $A/Ker\, h$ is isomorphic to $\mathbb{C}$, the complex field, which has no proper ideals, hence there are no intermediate ideals between $Ker\, h$ and A. Another property of $Ker\, h$ is *regularity*: there exists an element u in A, which is a *relative identity* with respect to $Ker\, h$, that is, $ux - x$ belongs to $Ker\, h$ for all x in A. Indeed, any u with $h(u) = 1$ satisfies this property. In general, an ideal is called *regular*, if there exists such a relative identity with respect to it, therefore, in a commutative algebra with identity any ideal is regular.

Hence to each h in H there corresponds a regular maximal ideal M_h, which is the kernel of h. It is easy to see that this correspondence is one-to-one, that is, different multiplicative linear functionals cannot have the same kernel. This is slightly different from the well-known situation in linear space theory, where two linear functionals have the same kernel if and only if one of them is a nonzero constant multiple of the other. Here the constant is necessarily 1. For a bijective correspondence between multiplicative linear functionals and regular maximal ideals one needs a statement as follows: any regular maximal ideal is the kernel of some multiplicative linear functional. In any case a regular maximal ideal M is the kernel of the natural homomorphism $A \mapsto A/M$, and by the maximality of M the factor algebra A/M is a field. The question is: in which case can A/M be identified with $\mathbb{C}$? The following theorem helps us.

Theorem 1.1. *(Gelfand–Mazur) Any normed field is isometrically isomorphic to the complex field.*

Consequently, it is useful to assume that A is a *normed algebra*. But, in general, this is not enough to assure that A/M is a normed field, because the standard technique, taking the norm from A onto A/M, works only if M is closed. Our problem is: in which case is any regular maximal ideal of A necessarily closed? To answer this question we need the concept of *adverse*. An adverse of the element x in A is the element y, if $xy - x - y = 0$. Heuristically $y = x \cdot (x-1)^{-1}$, but in general there is no identity and inverse. The adverse plays the role of an inverse in the lack of an identity. The following theorem holds.

Theorem 1.2. *In a commutative normed algebra every regular maximal ideal is closed if and only if every element x with $\|x\| < 1$ has an adverse.*

If the series converges, then $y = -\sum_{k=1}^{\infty} x^k$ is an adverse of x. On the other hand, for $\|x\| < 1$ this series is absolutely convergent. What we need is that any absolutely convergent series is convergent, which is the case exactly if A is a *Banach space*. Hence from now on it is quite reasonable to assume that A is a commutative *Banach algebra*. This is the natural setting for the theory of Gelfand transformation.

Suppose now that H consists of all continuous multiplicative linear functionals of A. The correspondence $h \mapsto Ker\, h$ is bijective, since every regular maximal ideal is closed, thus it is the kernel of a continuous multiplicative linear functional. Hence this correspondence identifies H with the space of all regular maximal ideals of A, with the *maximal ideal space* of A, which we denote by Δ. The maximal ideal space of A can be considered as the dual object of A.

Returning to the Gelfand transformation, for its bijectivity it is necessary and sufficient, that Δ is separating for A. In the language of functionals this means that there are sufficiently many continuous multiplicative linear functionals to separate the elements of A. In the language of maximal ideals this means that there exist no nonzero elements in A which belong to all the regular maximal ideals, that is, the intersection of all regular maximal ideals is zero. In general this intersection is called the *radical* of the algebra, and if it is zero, then A is called *semi-simple*. Of course, in general, it is a nontrivial problem to decide if a given commutative Banach algebra is semi-simple.

Usually Δ is given the *weak topology* induced by $\widehat{A}$, that is, the weakest topology with respect to which all Gelfand transforms are continuous. This is the same as the *weak*-topology* for Δ, if Δ is considered as a subset of A^*, the dual of A as a Banach space. It is easy to see that Δ is a subset of the *unit ball* in A^*, hence its closure is compact in the weak*-topology, by the Banach–Alaoglu Theorem. If A has an identity, then Δ itself is closed, hence it is compact. In general, Δ is *locally compact* and Haussdorff. If Δ is not compact, then all functions in $\widehat{A}$ vanish at infinity, which means that for any x in A and for any $\varepsilon > 0$ there exists a compact subset K of Δ such that $|\hat{x}(h)| < \varepsilon$ for any h not in K.

Summarizing these results, if A is a commutative Banach algebra and Δ is the locally compact Haussdorff space of all regular maximal ideals of A, then the Gelfand transformation is a norm-decreasing algebra homomorphism of A into the function algebra $\widehat{A}$, which consists of some continuous complex-valued functions defined on Δ and vanishing at infinity. The central problem of the theory of commutative Banach algebras is to find conditions which assure that the Gelfand transformation is one-to-one (the algebra is semi-simple) and is onto the space $\mathcal{C}_0(\Delta)$, the space of all continuous complex functions on Δ vanishing at infinity. If so, then the original algebra can be considered as the algebra of all continuous complex-valued functions defined on a locally compact Haussdorff space, and vanishing at infinity.

Here we present a very simple, but important example. Let X be a compact Haussdorff space and let $A = \mathcal{C}(X)$, the set of all continuous complex valued functions on X, which is a commutative Banach algebra with identity, if it is equipped with the sup-norm. In this algebra any maximal ideal is regular. For any p in X the set

$$M_p = \{f \, : \, f(p) = 0\}$$

is a closed ideal in $\mathcal{C}(X)$. If I is any proper ideal in $\mathcal{C}(X)$, then there exists a p in X for which M_p contains I. Indeed, if for any p in X there exists an f_p in I with $f_p(p) \neq 0$, then $|f_p|^2 = f_p \overline{f_p}$ belongs to I, $|f_p|^2 \geq 0$ and $|f_p|^2 > 0$ in a neighborhood of p. By compactness we get a continuous function f in I which is positive on X. Then $\frac{1}{f}$ is continuous and $1 = f\frac{1}{f}$ belongs to I, which contradicts the fact that I is a proper ideal.

It follows that the maximal ideals are exactly the ones of the form M_p with some p in X. If $p \neq q$, then by Uryshon's Lemma there exists an f in $\mathcal{C}(X)$ with $f(p) \neq 0$, $f(q) = 0$, and hence f is not in M_p, but f is in M_q. This means that the correspondence $p \mapsto M_p$ is bijective, and set-theoretically Δ can be identified with X. For this correspondence to be a homeomorphism, it is necessary and sufficient that the topology of X is identical with the one defined by $\mathcal{C}(X)$. But this is just the complete regularity of X, a consequence of compactness, that is, the topology of X is completely determined by the continuous complex valued functions on it.

The Gelfand transform for any f in $\mathcal{C}(X)$ is

$$\hat{f}(M_p) = h_p(f) = f(p),$$

that is, as Δ is identified with X, $\hat{f}$ can be identified with f.

Similar arguments can be used if X is a locally compact Haussdorff space and $A = \mathcal{C}_0(X)$.

For the proofs of the theorems in this section, for more details and for further references see [36].

2

Spectral synthesis and spectral analysis

2.1 The basic problems of spectral analysis and spectral synthesis

In the case of the ideals of $\mathcal{C}(X)$ we have seen that any proper ideal is included in a maximal ideal. We can prove the same for any proper regular ideal in any commutative algebra.

Theorem 2.1. *In a commutative algebra any proper regular ideal is included in a regular maximal ideal.*

Proof. Let u be a relative identity with respect to the given proper regular ideal I and consider the family of all proper ideals containing I. These are all regular, the family is partially ordered by inclusion and satisfies the condition of Zorn's Lemma, hence it has a maximal element.

Now we can formulate the basic problem of *spectral analysis*: in a commutative Banach algebra, is every proper closed ideal included in a regular maximal ideal? The answer is "yes" in $\mathcal{C}(X)$, where X is compact Haussdorff, but "no" in general.

A basic theorem in this context is the following. For the proof see [36]. We remark, that a *regular Banach algebra* is commutative, further for any closed set C of Δ and for any M not included in C there exists a function $\widehat{x}$ in $\widehat{A}$ with $\widehat{x}(M) \neq 0$ and $\widehat{x} = 0$ on C.

Theorem 2.2. *(Wiener Tauberian Theorem) Let A be a regular semi-simple Banach algebra with the property that the set of elements x in A such that $\widehat{x}$ has compact support is dense in A. Then every proper closed ideal of A is included in a regular maximal ideal.*

7

The formulation of the basic problem of *spectral synthesis* is the following: is it true that every proper closed ideal in a commutative Banach algebra is the intersection of all regular maximal ideals in which it is contained?

The answer is again "yes" for $C(X)$, where X is compact Haussdorff, and "no" for the general case. For let I be a proper closed ideal in $C(X)$ and consider the intersection of all maximal ideals containing I:

$$\bigcap_{I \subseteq M_p} M_p = \{f : f(p) = 0 \text{ if } I \subseteq M_p\}.$$

Let further

$$C = \{p : I \subseteq M_p\} = \{p : f(p) = 0 \text{ for all } f \text{ in } I\}.$$

Then C is closed, and the intersection of all maximal ideals containing I is

$$I_C = \bigcap_{I \subseteq M_p} M_p = \{f : f(p) = 0 \quad \text{for all} \quad p \quad \text{in} \quad C\}.$$

Let X_1 denote the complement of C in X, then X_1 is locally compact. The restrictions of the functions in I_C to X_1 form the algebra of functions vanishing at infinity in $C(X_1)$. Then I is closed in I_C. By Uryshon's Lemma for any $p \neq q$ in X_1 there exists an f in $C(X)$ with $f(p) = 1$, $f(q) = 0$ and $f = 0$ on C, further there exists a function g in I such that $g(p) \neq 0$. We infer that gf belongs to I, $gf(p) \neq 0$ and $gf(q) = 0$, that is, I is dense in I_C, by the Stone–Weierstrass Theorem. We conclude that $I = I_C$.

For further references on the results in this section see [2], [24], [36].

2.2 Spectral analysis and synthesis on $L^1(G)$

Now we specialize our setting by restricting ourselves to the algebra $L^1(G)$, where G is a locally compact Abelian group and multiplication is defined by convolution. With the L^1-norm this is a commutative Banach algebra which has an identity if and only if G is discrete. For the characterization of the maximal ideal space of $L^1(G)$, we know that for any M in Δ there exists a multiplicative linear functional on $L^1(G)$ with kernel M. As this functional is continuous and linear on $L^1(G)$ and the dual of $L^1(G)$ can be identified with $L^\infty(G)$, this functional can be uniquely represented by an element α_M of $L^\infty(G)$ in the form

$$\widehat{\varphi}(M) = \int \varphi(x)\overline{\alpha}_m(x)\, dx. \tag{2.1}$$

From the multiplicativity of α_M it follows that

$$\alpha_M(x+y) = \alpha_M(x)\alpha_M(y)$$

holds almost everywhere on $G \times G$. By standard methods (see e.g. Theorem 5.10. in [66], p.51.) it follows that α_M is almost everywhere equal to a continuous function with absolute value 1 satisfying the above functional equation for all x, y in G. This means that α_M can be identified with a function, which we call a *character* of G. Thus the maximal ideal space of $L^1(G)$ can be identified with the set $\widehat{G}$ of all characters of G. Through this identification we know that $\widehat{G}$ is locally compact in the weak*-topology of $L^\infty(G) = L^1(G)^*$. Further, it follows easily, that $\widehat{G}$ is a locally compact Abelian group, multiplication being defined as the pointwise multiplication of functions. We call $\widehat{G}$ the *dual* of G. It follows that $\widehat{G}$ is compact if and only if G is discrete. In addition, the Gelfand transformation has the form given in (2.1) for any φ in $L^1(G)$, where α_M is the character corresponding to the regular maximal ideal M. The function $\widehat{\varphi} : \widehat{G} \to \mathbb{C}$ is called the *Fourier transform* of the function φ and the mapping $\varphi \mapsto \widehat{\varphi}$ is the *Fourier transformation*.

We summarize our knowledge on the Fourier transformation. For any locally compact Abelian group G the Fourier transformation $\varphi \mapsto \widehat{\varphi}$ is a norm-decreasing algebra homomorphism of $L^1(G)$ onto a set of continuous, vanishing at infinity, complex-valued functions, defined on the locally compact Abelian dual group $\widehat{G}$ of G. We note that the statement that the Fourier transforms of the functions in $L^1(G)$ are vanishing at infinity is called *Mercer's Theorem* in the case $G = \mathbb{T}$, and the *Riemann–Lebesgue Lemma* in the case $G = \mathbb{R}$.

Without going into the details of the duality theory of locally compact Abelian groups, we state here some consequences of the famous *Duality Theorem of Pontryagin*:

i) The dual of the dual of G is G.
ii) The dual of a discrete group is compact, the dual of a compact group is discrete.
iii) The dual of $\mathbb{Z}$ is $\mathbb{T}$, the dual of $\mathbb{T}$ is $\mathbb{Z}$, and the dual of $\mathbb{R}$ is $\mathbb{R}$.

We state here also some basic theorems on the Fourier transformation. We recall that a function p in $L^\infty(G)$ is called *positive definite* if the corresponding linear functional on $L^1(G)$ is positive, that is, if

$$\int\int \varphi(x)\overline{\varphi(y)}p(x-y)\,dx\,dy \geq 0$$

for all φ in $L^1(G)$. For instance, any character is positive definite. It follows that convex combinations of positive definite functions are positive definite again. The famous theorem of Bochner states that if we allow "continuous" convex combinations, that is, integrals with respect to positive measures, then all positive definite functions have this form.

Theorem 2.3. *(Bochner) The formula*

$$p(x) = \int \alpha(x) \, d\mu(\alpha)$$

sets up a norm-preserving isomorphism between the convex set of all finite positive Baire measures μ on $\widehat{G}$ and the convex set of all positive definite functions p in $L^\infty(G)$.

The following theorem is about the inverse of the Fourier transformation restricted to a special set concerning positive definite functions. We denote by $[L^1(G) \cap P]$ the closed subspace generated by the positive definite functions in $L^1(G)$.

Theorem 2.4. *(Inversion Theorem) If φ is in $[L^1(G) \cap P]$, then $\widehat{\varphi}$ is in $L^1(\widehat{G})$ and*

$$\varphi(x) = \int \widehat{\varphi}(\alpha) \, \alpha(x) \, d\alpha$$

for almost all x in G, where $d\alpha$ is the Haar measure on $\widehat{G}$, suitably normalized.

One more important theorem in this relation is the following.

Theorem 2.5. *(Plancherel) The Fourier transformation on $[L^1(G) \cap P]$ preserves scalar product and its L^2-closure is a unitary operator of $L^2(G)$ onto $L^2(\widehat{G})$.*

Returning to our original problem we have the following result.

Theorem 2.6. $L^1(G)$ *is semi-simple and regular.*

This means that if the functions whose Fourier transforms vanish off compact sets are dense in $L^1(G)$, then we can apply the Wiener Tauberian Theorem for $L^1(G)$. It is easy to check that this is the case, hence we can restate the Tauberian Theorem for $L^1(G)$.

Theorem 2.7. *(Tauberian Theorem) If G is a locally compact Abelian group, then every proper closed ideal of $L^1(G)$ is included in a regular maximal ideal.*

In other words, spectral analysis holds for the proper closed ideals of $L^1(G)$.

For any y in G the symbol τ_y denotes the *translation* by y, which maps the function f onto $\tau_y f$, defined by $(\tau_y f)(x) = f(x + y)$ for any x in G. The closed subspace generated by the translates of f is denoted by $\tau(f)$. A set of functions on G is called *translation invariant* if it contains all translates of each of its elements. The above theorem implies the following approximation theorem of Wiener.

Theorem 2.8. *(Approximation Theorem of Wiener) If G is a locally compact Abelian group and the Fourier transform of a function in $L^1(G)$ never vanishes, then its translates generate $L^1(G)$.*

The proof of this theorem depends on the fact that the closed ideals of $L^1(G)$ are just its closed translation invariant subspaces. Having this in mind, by the Tauberian Theorem, φ belongs to no regular maximal ideal. Hence the closed ideal generated by φ, that is, the closed subspace generated by the translates of φ, is the whole of $L^1(G)$.

This is a typical spectral analysis theorem: if the translates of φ do not generate $L^1(G)$, then the Fourier transform of φ vanishes somewhere.

Here we state two famous consequences of the abstract Wiener Tauberian Theorem.

Theorem 2.9. *(Generalized Wiener Tauberian Theorem) Let G be a locally compact Abelian group which is not compact and let f be in $L^\infty(G)$. If there exists a function ψ in $L^1(G)$ with non-vanishing Fourier transform for which $f * \psi$ vanishes at infinity, then $f * \varphi$ vanishes at infinity for every φ in $L^1(G)$.*

Proof. The set I of all functions φ in $L^1(G)$ for which $f * \varphi$ vanishes at infinity is a closed translation invariant subspace, that is, a closed ideal, and since it includes ψ, it is not included in any regular maximal ideal. Hence $I = L^1(G)$.

Theorem 2.10. *(Classical Wiener Tauberian Theorem) Let f be a function in $L^\infty(\mathbb{R})$. If there exists a function ψ in $L^1(\mathbb{R})$ with non-vanishing Fourier transform such that $f * \psi(x) \to 0$ as $x \to +\infty$, then $f * \varphi(x) \to 0$ as $x \to +\infty$ for all φ in $L^1(\mathbb{R})$.*

The proofs of the above theorems can be found in [36]. For further references see also [2], [23] and [24].

2.3 Spectral analysis and synthesis on $L^\infty(G)$

As $L^\infty(G)$ is the dual of $L^1(G)$, several problems on $L^1(G)$ can be formulated in the dual language, as problems in $L^\infty(G)$. From now on we always suppose that $L^\infty(G)$ is equipped with the weak*-topology, and all topological concepts in $L^\infty(G)$ relate to this topology, if the opposite is not explicitly stated. From the general dual space theory we know the notion of the *annihilator*. If I is a subspace of $L^1(G)$, then let

$$I^\perp = \{f : f \in L^\infty(G), \langle f, \varphi \rangle = 0 \quad \text{for all} \quad \varphi \in I\},$$

and if V is a subspace of $L^\infty(G)$, then let

$$V^\perp = \{\varphi : \varphi \in L^1(G), \langle f, \varphi \rangle = 0 \quad \text{for all} \quad f \in V\}.$$

Then $I^\perp$, $V^\perp$ are closed subspaces in $L^\infty(G)$ and $L^1(G)$, respectively, and from the Banach space theory we know that $(I^\perp)^\perp = I$ holds for any closed subspace I in $L^1(G)$. As the dual of $L^\infty(G)$ with respect to the weak*-topology is just $L^1(G)$, we have also $(V^\perp)^\perp = V$ for any closed subspace V in $L^\infty(G)$. Hence we have two one-to-one correspondences between the closed subspaces of $L^1(G)$ and $L^\infty(G)$, which are inverse to each other. We show that at this correspondence, ideals of $L^1(G)$ correspond to translation invariant subspaces of $L^\infty(G)$.

Theorem 2.11. *Let V be a closed subspace in $L^\infty(G)$. Then $V^\perp$ is a closed ideal if and only if V is translation invariant.*

Proof. Suppose, that V is translation invariant. If φ is in $V^\perp$, ψ is in $L^1(G)$ and f is in V, then $\tau_y f$ belongs to V for all y in G, hence

$$\langle \tau_y f, \varphi \rangle = 0,$$

which implies

$$\langle f, \varphi * \psi \rangle = \int \int \overline{f}(x) \varphi(x - y) \psi(y) \, dy \, dx$$
$$= \int \int \overline{f}(x + y) \varphi(x) \psi(y) \, dx \, dy = 0,$$

that is, $\varphi * \psi$ belongs to $V^\perp$, hence $V^\perp$ is a closed ideal.

Conversely, if $V^\perp$ is a closed ideal in $L^1(G)$, then for any φ in $V^\perp$, f in V and ψ in $L^1(G)$ we have

$$0 = \langle f, \varphi * \psi \rangle = \int \int \overline{f}(x) \varphi(x - y) \psi(y) \, dy \, dx$$
$$= \int \int \overline{f}(x + y) \varphi(x) \psi(y) \, dx \, dy,$$

that is, the function $y \mapsto \langle \tau_y f, \varphi \rangle$ annihilates $L^1(G)$, hence $\langle \tau_y f, \varphi \rangle$ vanishes, and by the Hahn–Banach Theorem, $\tau_y f$ is in V.

Hence we have a one-to-one correspondence between the closed ideals of $L^1(G)$ and the closed translation invariant subspaces of $L^\infty(G)$. First we characterize the closed ideals of $L^1(G)$. For doing so we need the approximate identity technique, which is based on the following theorem of L. Fejér.

Theorem 2.12. *(Fejér) For any neighborhood U of the identity in G there exists a nonnegative function e_U in $L^1(G)$, which vanishes off U and satisfies $\int e_U = 1$, further $e_U * \varphi \to \varphi$ in $L^1(G)$ for all φ in $L^1(G)$. If φ is continuous, then $e_U * \varphi$ converges uniformly.*

The above convergence is meant in the sense of net-convergence along the partially ordered set of all neighborhoods of the identity in G. Such a net $\{e_U\}$ is called an *approximate identity*. Using this theorem the next one follows easily.

Theorem 2.13. *A closed subset I in $L^1(G)$ is an ideal if and only if it is a translation invariant subspace.*

Proof. Let I be a closed ideal in $L^1(G)$ and let e_U be an approximate identity. If φ is in I, then $\tau_x e_U * \varphi$ belongs to I. But $\tau_x e_U * \varphi = e_U * \tau_x \varphi \to \tau_x \varphi$, hence $\tau_x \varphi$ is in I. Thus every closed ideal is a translation invariant subspace.

Now let I be a closed translation invariant subspace. Then $I = (I^\perp)^\perp$, that is, if φ is in $L^1(G)$, then φ belongs to I if and only if $\langle f, \varphi \rangle = 0$ for every f in $I^\perp$. But if ψ is in $L^1(G)$, φ is in I and f is in $I^\perp$, then

$$\langle f, \varphi * \psi \rangle = \int \int \overline{f}(x) \varphi(x - y) \psi(y) \, dy \, dx$$

$$= \int \int \tau_{-y} \varphi(x) \overline{f}(x) \psi(y) \, dx \, dy = 0,$$

which proves that $\varphi * \psi$ is in I, that is, I is an ideal.

Now we try to translate our basic theorems on spectral analysis and spectral synthesis into the language of $L^\infty(G)$. What does it mean for $I^\perp$ that I is contained in a regular maximal ideal in $L^1(G)$? Or simply : what does it mean for $I^\perp$ that I is a regular maximal ideal in $L^1(G)$? Then $I^\perp$ has no proper closed translation invariant subspace. On the other hand, as I is the kernel of some multiplicative linear functional, that is,

$$I = \{\varphi : \widehat{\varphi}(\gamma) = 0\}$$

for some γ in $\widehat{G}$, hence it follows that γ belongs to $I^\perp$ and then $I^\perp$ is equal to the one-dimensional subspace generated by γ. That is, closed proper minimal invariant subspaces are just the one-dimensional subspaces, generated by characters. In other words, I is a regular maximal ideal in $L^1(G)$ if and only if $I^\perp$ is a one-dimensional subspace of $L^\infty(G)$ generated by a character.

What does spectral analysis for a proper closed ideal I of $L^1(G)$ imply for $I^\perp$? The Tauberian Theorem can be reformulated.

Theorem 2.14. *Any proper closed translation invariant subspace of $L^\infty(G)$ contains a character.*

This is the first point where we can give an example for the possible applications of the results of spectral analysis in the theory of functional equations. Namely, we can solve d'Alembert's classical equation in $L^\infty(G)$.

Theorem 2.15. *Let G be a locally compact Abelian group and $f : G \to \mathbb{C}$ a nonzero bounded and measurable function satisfying*

$$f(x+y) + f(x-y) = 2f(x)f(y) \tag{2.2}$$

for all y and for almost all x in G. Then f is continuous and

$$f(y) = \frac{1}{2}\left(\gamma(y) + \gamma(-y)\right) \tag{2.3}$$

holds for all y in G with some character γ of G.

Proof. As f is nonzero, $\tau(f)$ contains a character γ. But any g in $\tau(f)$ satisfies

$$g(x+y) + g(x-y) = 2g(x)f(y)$$

for all y and for almost all x in G, hence we have for γ,

$$\gamma(x+y) + \gamma(x-y) = 2\gamma(x)f(y)$$

for all x, y in G. As the left-hand side is continuous in y, hence f is continuous, and dividing by $\gamma(x) \neq 0$ we get our statement.

We remark that the classical way of solving d'Alembert's functional equation is highly nontrivial, even in the real case. The interested reader should consult [1].

The main problem of spectral analysis for $L^1(G)$ was: is every proper closed ideal of $L^1(G)$ contained in a regular maximal ideal? This has been reformulated for $L^\infty(G)$: does every proper closed translation invariant subspace of $L^\infty(G)$ contain a character? The answer is "yes", due to the Wiener Tauberian Theorem. "Spectral analysis of a proper closed translation invariant subspace" of $L^\infty(G)$ means to determine the characters in this subspace. If we take an f in $L^\infty(G)$, then "spectral analysis of f" means to determine the characters of $\tau(f)$. These are the candidates for taking part in the "spectral synthesis of f", that is, in the reconstruction process of f from characters. The set of all characters in a closed translation invariant subspace V of $L^\infty(G)$ is called the *spectrum* of V. Accordingly, the spectrum of $\tau(f)$ is called the spectrum of f. Notation: $sp\, V$ or $sp\, f$. It is easy to see that the spectrum of V is identical with the set of common zeros of the Fourier transforms of the φ's in $L^1(G)$, annihilating V. In particular, the character γ belongs to the spectrum of f if and only if $\widehat{\varphi}(\gamma) = 0$ for all φ in $L^1(G)$ with $f^* * \varphi = 0$, where f^* is defined by $f^*(x) = \overline{f}(-x)$.

Let us, for instance, determine the spectrum of a character γ in $L^\infty(G)$. As $\tau(\gamma)$ consists of the scalar multiples of γ, the only character in $\tau(\gamma)$ is γ, hence the spectrum of $\tau(\gamma)$ is $\{\gamma\}$. Similarly, if f is a *trigonometric polynomial*, that is, if it has the form $\sum_{i=1}^n c_i\gamma_i$, where $\gamma_1, \gamma_2, \ldots, \gamma_n$ are characters and

$c_1, c_2, \ldots, c_n$ are complex numbers, then the spectrum of f is $\{\gamma_1, \gamma_2, \ldots, \gamma_n\}$. We repeat it again: the elements of the spectrum of f in $L^\infty(G)$ are those characters which are weak*-limits of nets formed by linear combinations of translates of f. Practically these can be computed, if we determine all common zeros of the Fourier transforms of the elements in $\tau(f)^\perp$, that is, of the functions φ in $L^1(G)$ with $f^* * \varphi = 0$. Suppose, for instance, that f is in $L^1(G) \cap L^\infty(G)$. Then $f^* * \varphi = 0$ implies $\widehat{f^*} \widehat{\varphi} = 0$, hence by $\widehat{f^*} = \widehat{\bar{f}}$ the spectrum of f is closely related to the support of $\widehat{f}$. This is quite natural in the light of the Inversion Theorem.

By the definition of the spectrum of f in $L^\infty(G)$, it follows that if $f^* * \varphi = 0$ for some φ in $L^1(G)$, then $\widehat{\varphi} = 0$ on $sp\, f$. The converse holds under a stronger assumption.

Theorem 2.16. *If φ is in $L^1(G)$ and $\widehat{\varphi}$ vanishes on a neighborhood of $sp\, f$ for some f in $L^\infty(G)$, then $f^* * \varphi = 0$.*

The question, whether $f^* * \varphi = 0$ on the hypothesis merely that $\widehat{\varphi}$ vanishes on $sp\, f$, was the celebrated problem of A. Beurling (see [4], [5], [23]): If φ is in $L^1(G)$ and $\widehat{\varphi}$ vanishes on $sp\, f$ for some f in $L^\infty(G)$, does it follow $f^* * \varphi = 0$?

We show that the Beurling problem has an affirmative answer if and only if f is the weak*-limit of trigonometric polynomials in $sp\, f$, or equivalently, if for $\tau(f)^\perp$ in $L^1(G)$ spectral synthesis holds. First we show the equivalence of the two latter statements.

Theorem 2.17. *For a proper closed ideal I in $L^1(G)$, spectral synthesis holds if and only if the trigonometric polynomials of $I^\perp$ are dense in $I^\perp$.*

Proof. Let J be the closure in $I^\perp$ of the subspace spanned by its trigonometric polynomials. Obviously, J is a closed translation invariant subspace of $I^\perp$, which is nonzero, by spectral analysis. Then we have

$$I = (I^\perp)^\perp \subseteq J^\perp,$$

and here $J^\perp$ is a proper closed ideal in $L^1(G)$. If $M = M_\gamma$ is any regular maximal ideal, containing I, then γ is in $I^\perp$, hence γ is in J, and $J^\perp$ is a subset of $M = M_\gamma$. It means that any regular maximal ideal which contains I contains $J^\perp$ as well. Hence, if spectral synthesis holds for I, then $I^\perp = J$.

Conversely, let $I^\perp = J$. We observe that φ belongs to $J^\perp$ if and only if $\widehat{\varphi}(\gamma) = 0$ for all γ in $sp\, J$. That is, φ belongs to $J^\perp$ if and only if φ is in M_γ for all γ in $sp\, J$. In other words, φ belongs to $J^\perp$ if and only if φ is in M_γ for all γ with $J^\perp$ is a subset of M_γ, that is, φ belongs to $J^\perp$ if and only if φ is in $\bigcap_{J^\perp \subseteq M_\gamma} M_\gamma$. This shows that $J^\perp$ is the intersection of all regular maximal ideals containing it, that is, spectral synthesis holds for $I = (I^\perp)^\perp = J^\perp$.

Now we show that a positive answer to the Beurling question can be given if and only if, for $\tau(f)^\perp$ in $L^1(G)$, spectral synthesis holds. Indeed, this latter statement is equivalent to the relation $\tau(f)^\perp = \bigcap_{\tau(f)^\perp \subseteq M_\gamma} M_\gamma$. But $\tau(f)^\perp \subseteq M_\gamma$ means that any φ in $L^1(G)$ with $f^* * \varphi = 0$ satisfies $\widehat{\varphi}(\gamma) = 0$, that is $\tau(f)^\perp \subseteq M_\gamma$ if and only if γ is in $sp\, f$. Hence for $\tau(f)^\perp$ in $L^1(G)$ spectral synthesis holds if and only if $\tau(f)^\perp = \bigcap_{\gamma \in sp\, f} M_\gamma$, which is equivalent to the fact that for any φ in $L^1(G)$ $f^* * \varphi = 0$ holds if and only if $\widehat{\varphi}(\gamma) = 0$ for all γ in $sp\, f$, and this was to be proved.

We summarize what it means for $I^\perp$ in $L^\infty(G)$, that spectral synthesis holds for I in $L^1(G)$: this means precisely that the trigonometric polynomials of $I^\perp$ are weak*-dense in $I^\perp$. A negative solution for the Beurling problem in $\mathbb{R}^3$ was given by L. Schwartz in 1947 (see [56]), who presented a counterexample. From a previous theorem we know that any element of $\tau(f)$ can be approximated by trigonometric polynomials, taken from a neighborhood of $sp\, f$. It follows that if $\widehat{G}$ is discrete, then we have a positive solution for the Beurling problem. This is half of the following theorem (see [39]).

Theorem 2.18. *(Malliavin) Spectral synthesis holds for $L^1(G)$ if and only if G is compact.*

This means that all proper closed ideals of $L^1(G)$ have spectral synthesis if and only if G is compact. Nevertheless, in general, there may exist some special proper closed ideals in $L^1(G)$ for which spectral synthesis holds. The following theorem presents a case of this type.

Theorem 2.19. *(Primary Ideal Theorem) If the spectrum of a bounded measurable function has exactly one point, then the function is a constant multiple of a character.*

For any locally compact Abelian group this theorem is due to I. Kaplansky (see [30]), who used the structure theory of locally compact Abelian groups for his proof. Previously the theorem was proved for $G = \mathbb{R}$ by V. Ditkin (see [11]). An independent proof based on distribution theory was given by J. Riss (see [49]). Another proof for the general case which does not depend on the structure theory was given by H. Helson (see [22]). The theorem has its name, as the dual statement reads: a closed ideal of functions in $L^1(G)$ whose Fourier transforms have only one common zero necessarily contains all functions whose Fourier transforms vanish at that point. Ideals which are contained in precisely one regular maximal ideal are called *primary ideals*. We note that a simple analogue of the Primary Ideal Theorem in the case $\mathcal{C}(X)$, where X is a compact Haussdorff space, is the following: if the functions of a proper ideal in $\mathcal{C}(X)$ have only one common zero, then this ideal is just the maximal ideal of all functions vanishing at that point.

Extensions of the Primary Ideal Theorem on the real line are due to V. Ditkin, I. E. Segal, S. Mandelbrojt and S. Agmon ([11], [40], [57]). These are of the following type: if the boundary of the spectrum of an f in $L^\infty(G)$ is denumerable, or it does not contain any nonempty perfect set, then f is a limit of trigonometric polynomials on $sp\,f$.

2.4 Spectral analysis and synthesis on $\mathcal{C}(G)$

Our last formulation of the basic spectral problems was the following:

i) Does every proper weak*-closed invariant subspace of $L^\infty(G)$ contain a character?
ii) Are the trigonometric polynomials of every proper weak*-closed invariant subspace of $L^\infty(G)$ dense in this subspace?

This formulation depends on the fact that $L^1(G)$ is the dual of $L^\infty(G)$, if the latter is equipped with the weak*-topology. However, we can consider this problem from a more general point of view. Namely, the dual pair $\left(L^\infty(G), L^1(G)\right)$ can be replaced by several other dual pairs $\left(\mathcal{F}(G), \mathcal{F}(G)^*\right)$, where $\mathcal{F}(G)$ is a given translation invariant topological vector space of functions on the locally compact Abelian group G and $\mathcal{F}(G)^*$ is its dual. Then the two basic problems will have the following form:

i) Does every proper closed invariant subspace of $\mathcal{F}(G)$ contain "minimal" translation invariant subspaces?
ii) Do the "minimal" translation invariant subspaces of every proper closed translation invariant subspace of $\mathcal{F}(G)$ "generate" this subspace?

Of course, a precise meaning must be given to the words "minimal" and "generate". We have seen that if a convolution in $\mathcal{F}(G)^*$ can be defined, then the "minimal" translation invariant subspaces of $\mathcal{F}(G)$ relate somehow to the multiplicative linear functionals of the algebra $\mathcal{F}(G)^*$. For instance, if G is a locally compact Abelian group and $\mathcal{F}(G) = \mathcal{C}(G)$, the set of all continuous, complex valued functions on G equipped with the topology of uniform convergence on compact sets, then $\mathcal{F}(G)^*$ is the set of all compactly supported complex Radon measures on G, which is a commutative algebra with identity with the convolution of measures. In order to understand the situation better, we present here the simple example of $G = \mathbb{Z}$ with the discrete topology, where $\mathcal{F}(\mathbb{Z}) = \mathcal{C}(\mathbb{Z})$ is the set of all complex valued functions on $\mathbb{Z}$. The above topology on $\mathcal{C}(\mathbb{Z})$ is the topology of pointwise convergence, and $\mathcal{C}(\mathbb{Z})^*$ is the set of all finitely supported complex measures on $\mathbb{Z}$, which can also be realized as the set of all finitely supported complex functions on $\mathbb{Z}$. The maximal ideal space of $\mathcal{C}(\mathbb{Z})^*$ can be identified with the set of all nonzero complex numbers in the following manner: any maximal ideal of $\mathcal{C}(\mathbb{Z})^*$ is the kernel of a multiplicative linear functional h_λ with a complex nonzero λ of the form

$$h_\lambda(\mu) = \int \lambda^{-n}\, d\mu(n) = \sum_{k \in supp\, \mu} \mu(k)\lambda^{-k}\,,$$

where $\mu = \sum_{k \in supp\, \mu} \mu(k)\delta_k$ and δ_k denotes the Dirac measure concentrated at the point k for any k in $\mathbb{Z}$. Obviously any h_λ of this form is a multiplicative linear functional on $\mathcal{C}(\mathbb{Z})^*$. Conversely, suppose that h is a multiplicative linear functional on $\mathcal{C}(\mathbb{Z})^*$. As $\delta_{n+1} = \delta_n * \delta_1$ holds for any n in $\mathbb{Z}$, hence $h(\delta_n) = \lambda^{-n}$ with $\lambda = h(\delta_1)^{-1}$, which implies that $h = h_\lambda$, and the correspondence between h_λ and λ is one-to-one. As $\widehat{\mathbb{Z}} = \mathbb{T}$, we can see that the characters of $\mathbb{Z}$ are not sufficient to represent all multiplicative linear functionals of $\mathcal{C}(\mathbb{Z})^*$: the function $n \mapsto \lambda^n$ is a character of $\mathbb{Z}$ if and only if $|\lambda| = 1$. But any nonzero complex λ defines a function m_λ by the formula

$$m_\lambda(n) = \lambda^n$$

on $\mathbb{Z}$, which enjoys the fundamental property of characters

$$m_\lambda(n + k) = m_\lambda(n)m_\lambda(k)\,,$$

although it is not necessarily bounded. It is called a *generalized character* or an *exponential* of $\mathbb{Z}$. Hence the maximal ideal space of $\mathcal{C}(\mathbb{Z})^*$ can be identified with the set of all exponentials of $\mathbb{Z}$. It seems to be reasonable to modify our basic problem on spectral analysis for $\mathcal{C}(\mathbb{Z})$: does every proper closed translation subspace of $\mathcal{C}(\mathbb{Z})$ contain an exponential?

Now we consider any nonzero μ of the form

$$\mu = \sum_{k=0}^{N} c_k \delta_{-k}$$

in $\mathcal{C}(\mathbb{Z})^*$. The function f^* is an element of the annihilator of the ideal generated by μ if and only if f is a solution of the difference equation

$$f * \mu(n) = \sum_{k=0}^{N} c_k f(n + k) = 0$$

for all n in $\mathbb{Z}$, where we may suppose that $c_N = 1$. It is known from the theory of difference equations that in this case f is a linear combination of solutions of the form

$$n \mapsto n^j \lambda^n\,,$$

called *exponential monomials*, where λ is a characteristic root of the equation, that is $\sum_{k=0}^{N} c_k \lambda^k = 0$, and j is any nonnegative integer, smaller than the multiplicity of the root λ. This means that the annihilator of μ contains exponentials (not necessarily characters!), but the linear combinations of these exponentials are not necessarily dense. They form a dense set only if the characteristic polynomial of the difference equation has merely simple roots.

Hence a reasonable modification of our spectral synthesis problem for $\mathcal{C}(\mathbb{Z})$ is: are the linear combinations of exponential monomials belonging to a proper closed translation subspace of $\mathcal{C}(\mathbb{Z})$ dense in this subspace?

We observe that the characteristic polynomial of the difference equation is equal to

$$\widehat{\mu}(\lambda) = \int \lambda^{-n}\, d\mu(n),$$

which is a natural extension of the Fourier transform of μ from the set of all characters to the set of all exponentials. Then the closed ideal generated by μ is not the intersection of maximal ideals, but of ideals of the form

$$\{\mu : \widehat{\mu}(\lambda) = \widehat{\mu}'(\lambda) = \cdots = \widehat{\mu}^{(j)}(\lambda) = 0\}$$

for some complex λ and nonnegative integer j. These ideals have the property of being contained in exactly one maximal ideal, hence they are primary ideals.

For different choices of G and $\mathcal{C}(G)$ these two questions have been dealt with by several authors and have been answered either in the positive or in the negative. We would like to deal mostly with the case $\mathcal{F}(G) = \mathcal{C}(G)$, the space of all continuous complex valued functions on the locally compact topological group G. As indicated above, this space, equipped with the pointwise operations and with the topology of uniform convergence on compact sets, is a locally convex topological vector space. Its dual can be identified with the space of all compactly supported complex Radon measures on G, equipped with the weak*-topology and denoted by $\mathcal{M}_c(G)$. The pairing between $\mathcal{C}(G)$ and $\mathcal{M}_c(G)$ is given by

$$\langle f, \mu \rangle = \int \overline{f}(x)\, d\mu(x)$$

for all f in $\mathcal{C}(G)$ and μ in $\mathcal{M}_c(G)$. The convolution between $\mathcal{C}(G)$ and $\mathcal{M}_c(G)$, further between $\mathcal{M}_c(G)$ and $\mathcal{M}_c(G)$, is defined in the usual way.

A continuous homomorphism of G into the multiplicative group of nonzero complex numbers is called an *exponential* or *generalized character*. All exponentials on G form a group with respect to pointwise multiplication. This group, equipped with the topology of uniform convergence on compact sets, is a locally compact Abelian group, which we call the *generalized character group* or *generalized dual* of G. Notation: $\widetilde{G}$. In general $\widetilde{G} \neq G$, but if G is compact, then $\widetilde{G} = \widehat{G}$.

The continuous homomorphisms of G into the additive group of complex numbers are called *additive* functions. A function $x \mapsto P\big(a_1(x), \ldots, a_n(x)\big)$ on G is called a *polynomial*, if P is a complex polynomial in n variables and $a_1, a_2, \ldots, a_n$ are additive functions. A product of a polynomial and an exponential is called an *exponential monomial*, and linear combinations of exponential monomials are called *exponential polynomials*. Hence the general form of an exponential monomial is

$$x \mapsto P\big(a_1(x), a_2(x), \ldots, a_n(x)\big)m(x),$$

and of an exponential polynomial is

$$x \mapsto \sum_{i=1}^{N} P_i\big(a_1(x), a_2(x), \ldots, a_n(x)\big)m_i(x),$$

where $P, P_1, P_2, \ldots, P_N$ are complex polynomials in n variables, $a_1, a_2, \ldots, a_n$ are additive functions and $m_1, m_2, \ldots, m_N$ are exponentials.

The Fourier transform of a compactly supported complex Radon measure μ on G has a natural extension from $\check{G}$ to $\widetilde{G}$, which is called the *Fourier–Laplace transform* of μ and is given by

$$\widehat{\mu}(m) = \int m(-x)\, d\mu(x)$$

for each m in $\widetilde{G}$. The above formula can be rewritten in the simpler form $\widehat{\mu}(m) = \langle \overline{m^*}, \mu \rangle$.

A proper closed translation invariant subspace of $\mathcal{C}(G)$ will be called a *variety*. For any subset H of $\mathcal{C}(G)$ the *variety generated by* H is the smallest variety containing H and is denoted by $\tau(H)$. For $H = \{f\}$ we write $\tau(f)$ instead of $\tau(\{f\})$. If $\tau(f)$ is not the whole of $\mathcal{C}(G)$, then we call f *mean periodic*. This means that the linear combinations of the translates of f are not dense in $\mathcal{C}(G)$. If G is infinite, then any exponential polynomial on G is mean periodic.

The annihilators of any sets in $\mathcal{C}(G)$ and $\mathcal{M}_c(G)$ will be used and will be denoted similarly as before. Further, we have the respective theorem.

Theorem 2.20. *The annihilator of a closed subspace in $\mathcal{C}(G)$ is a closed ideal if and only if the subspace is a variety.*

Hence we have a one-to-one correspondence between the varieties of $\mathcal{C}(G)$ and the closed proper ideals of $\mathcal{M}_c(G)$ again, and the basic spectral problems can be formulated equivalently either in $\mathcal{C}(G)$ or in $\mathcal{M}(G)$:

i) Does every variety in $\mathcal{C}(G)$ contain an exponential? — or equivalently — is every proper closed ideal of $\mathcal{M}_c(G)$ contained in a maximal ideal?

ii) Are the linear combinations of all exponential monomials of a variety in $\mathcal{C}(G)$ dense in this variety? — or equivalently — is every proper closed ideal in $\mathcal{M}_c(G)$ the intersection of primary ideals?

The relation between the two problems on $L^\infty(G)$ is obvious: as any nonzero bounded exponential monomial is actually a character, hence spectral synthesis implies spectral analysis in $L^\infty(G)$. On the other hand, from results of [66] it follows that if an exponential monomial of the form pm belongs to a variety, where p is a nonzero polynomial and m is an exponential, then the exponential m also belongs to the same variety. This means that spectral synthesis for a variety implies spectral analysis for the variety, too.

By the theory of finite difference equations in our former example $G = \mathbb{Z}$ the answer is affirmative for both questions. The first classical result in this respect is due to L. Schwartz (see [55]).

Theorem 2.21. *(Schwartz) In $\mathcal{C}(\mathbb{R})$ any variety is the closed linear hull of the exponential monomials which are contained in it.*

In particular, any variety contains an exponential. Using this fact, we can give all continuous — not necessarily bounded — solutions of the functional equation of d'Alembert simply by repeating the argument used in the bounded case. Schwartz proved also the analogous result for the varieties of $\mathcal{E}(\mathbb{R})$, the space of infinitely differentiable functions on $\mathbb{R}$ with the usual topology ([55]).

An extension of this result for $\mathbb{Z}^n$ is due to M. Lefranc (see [35]).

Theorem 2.22. *(Lefranc) In $\mathcal{C}(\mathbb{Z}^n)$ any variety is the closed linear hull of the exponential monomials which are contained in it.*

Using this theorem one can prove the following result (see [70]).

Theorem 2.23. *Spectral synthesis holds for any finitely generated discrete Abelian group.*

As any finitely generated Abelian group is the homomorphic image of $\mathbb{Z}^n$ for some n, Theorem 2.23 is the consequence of the following general result.

Theorem 2.24. *If spectral synthesis holds for an Abelian group, then it holds for its homomorphic images, too.*

Proof. Suppose that G is an Abelian group, H is a homomorphic image of G and let $F : G \to H$ be a surjective homomorphism. If V is a variety in $\mathcal{C}(H)$, then we let
$$V_F = \{f \circ F : f \in V\}.$$
Using the surjectivity of F, a routine calculation shows that V_F is a variety in $\mathcal{C}(G)$. Let Φ be an exponential monomial in V_F of the form
$$\Phi(x) = P\big(A_1(x), A_2(x), \ldots, A_n(x)\big)M(x), \tag{2.4}$$

where $A_1, A_2, \ldots, A_n$ are linearly independent additive functions on G, M is an exponential on G, and P is a complex polynomial in n variables. By the Lemma on page 18. in [70] the exponential M is in V_F, too, hence $M = m \circ F$ holds for some m in V. If u, v are arbitrary in H, then $u = F(x)$ and $v = F(y)$ for some x, y in G, which implies

$$m(u + v) = m(F(x) + F(y)) = m(F(x + y)) = M(x + y) = M(x)M(y)$$
$$= m(F(x))m(F(y)) = m(u)m(v).$$

As m is never zero, hence m is an exponential in V. On the other hand, (2.4) implies that

$$q(x) = P(A_1(x), A_2(x), \ldots, A_n(x)) = p(F(x))$$

holds for each x in G with some function $p : H \to \mathbb{C}$. We show that p is a polynomial on H. Using the Newton Interpolation Formula and the Taylor Formula in several variables, it follows easily that the functions $A_1, A_2, \ldots, A_n$ can be expressed as a linear combination of some translates of q. On the other hand, if $F(x) = F(y)$ for some x, y in G, then $q(x + z) = q(y + z)$ holds for each z in G, hence $A_i(x) = A_i(y)$ for $i = 1, 2, \ldots, n$. It follows that we can define the functions $a_i : H \to \mathbb{C}$ for $i = 1, 2, \ldots, n$ by the equation

$$a_i(u) = A_i(F(x)),$$

where x is arbitrary in G with the property $F(x) = u$. Further, we see immediately that a_i is additive for $i = 1, 2, \ldots, n$. On the other hand,

$$p(u) = p(F(x)) = P(A_1(x), A_2(x), \ldots, A_n(x)) = P(a_1(u), a_2(u), \ldots, a_n(u))$$

holds for any u in H, hence p is a polynomial on H. This means that the exponential monomial Φ above has the form $\Phi = \varphi \circ F$ with some exponential monomial φ in V. Finally, it is straightforward to verify that if the exponential monomials span a dense subspace in V_F, then the corresponding exponential monomials span a dense subspace in V, so our proof is complete.

It turns out that the extension of the theorem of Schwartz for $\mathbb{R}^n$ is not possible if $n > 1$ (see [55]). In 1965 R. J. Elliot made an attempt to show that this extension was possible for any discrete Abelian group, but unfortunately there was a gap in the proof of his theorem (see [16]). In particular, this fact puts even the problem on spectral analysis for general discrete Abelian groups into a new setting. We will come back to this problem in Section 3.5.

Although spectral analysis and spectral synthesis do not hold for any variety in any locally compact Abelian group, there are some special varieties which have spectral synthesis. These special varieties are characterized by the property that their annihilator ideal is a *principal ideal*, which means that it is generated by a single measure. A classical result in this direction is due to B. Malgrange (see [38]).

Theorem 2.25. *(Malgrange) For any nonzero linear partial differential operator $P(D)$ in $\mathbb{R}^n$, the linear hull of the exponential monomial solutions of the partial differential equation $P(D)f = 0$ is dense in the set of all solutions.*

For $n = 1$ this theorem reduces to the well-known fact about homogeneous linear differential equations with constant coefficients: the solutions are linear combinations of exponential monomial solutions. Here the basic space is $\mathcal{E}(\mathbb{R}^n)$, the space of infinitely differentiable functions on $\mathbb{R}^n$ with the usual topology, and $\mathcal{E}(\mathbb{R}^n)^*$ is the space of all distributions with compact support. The annihilator ideal is generated by the distribution $P(D)\delta$.

L. Ehrenpreis extended the principal ideal technique to obtain the following theorem (see [14], [13]).

Theorem 2.26. *(Ehrenpreis) If the annihilator of a variety in $\mathcal{E}(\mathbb{C}^n)$ is a principal ideal, then the variety is the closed linear hull of the exponential monomials which are contained in it.*

The respective extension of this theorem to $\mathcal{C}(G)$, where G is a locally compact Abelian group is due to Elliot and J. E. Gilbert (see [15], [21], [20]).

Theorem 2.27. *(Elliot–Gilbert) If G is a locally compact Abelian group, then in $\mathcal{C}(G)$ any variety whose annihilator ideal is a principal ideal is the closed linear hull of the exponential monomials which are contained in it.*

Summarizing the results, we have spectral synthesis for any variety in $\mathcal{C}(\mathbb{R})$ and in $\mathcal{C}(G)$, where G is discrete and finitely generated, and we have restricted spectral synthesis, that is spectral synthesis for those varieties in $\mathcal{C}(G)$ whose annihilator ideal is principal, if G is any locally compact Abelian group. In the following chapter we show that spectral analysis holds for discrete Abelian torsion groups (see also [69]).

References: [11], [4], [57], [55], [56], [5], [30], [49], [40], [22] [36], [38], [13], [14], [35], [39], [24], [15], [16] [1], [20], [21], [2], [23], [66], [69], [70].

3

Spectral analysis and spectral synthesis on discrete Abelian groups

3.1 Spectral analysis on discrete Abelian torsion groups

Let G be an Abelian group. We say that G is a *torsion group* if every element of G has finite order. In other words, for every x in G there exists a positive integer n with $nx = 0$. Hence G is not a torsion group if and only if there exists an element of G which generates a subgroup isomorphic to $\mathbb{Z}$.

We shall make use of the following result. The proof can be found in [24].

Theorem 3.1. *Let G be an Abelian group, H a subgroup of G and D a divisible Abelian group. If $\varphi : H \to D$ is a homomorphism, then there exists a homomorphism $\Phi : G \to D$ which extends φ, that is, $\Phi(x) = \varphi(x)$ for each x in H.*

Theorem 3.2. *Let G be an Abelian group. Then G is a torsion group if and only if every nonzero exponential monomial on G is a constant multiple of a character.*

Proof. Suppose that G is a torsion group. Let $a : G \to \mathbb{C}$ be an additive function and $m : G \to \mathbb{C}$ an exponential function. For every x in G there exists a positive integer n with $nx = 0$ and hence

$$0 = a(nx) = na(x),$$

which implies $a(x) = 0$, meaning that every additive function on G is zero and every polynomial is constant. Further

$$1 = m(nx) = m(x)^n,$$

which implies $|m(x)| = 1$. This means that every exponential function on G is a character. We conclude that if G is a torsion group, then every nonzero exponential monomial on G is a constant multiple of a character.

Assume now that G is not a torsion group, that is, there exists an x_0 in G such that the cyclic group generated by x_0 is isomorphic to $\mathbb{Z}$. Let $\alpha \neq 0$ be a complex number with $|\alpha| \neq 1$ and define $\varphi(nx_0) = n\alpha$ for each integer n. Then φ is a homomorphism of the subgroup generated by x_0 into the additive group of complex numbers. As this latter group is divisible, by the previous theorem this homomorphism can be extended to a homomorphism $a : G \to \mathbb{C}$ of G into the additive group of complex numbers. By $a(x_0) = \varphi(x_0) = \alpha \neq 0$ we have that a is a nonzero additive function, that is, a nonzero exponential monomial on G, which is obviously not a character by $|\alpha| \neq 1$. The theorem is proved.

Now we show that if G is a discrete Abelian torsion group, then any variety in $\mathcal{C}(G)$ contains a character. This is the analogue of the Tauberian Theorem 2.7, meaning that spectral analysis holds for discrete Abelian torsion groups (see [69]). The proof depends on Theorem 2.23.

Theorem 3.3. *Let G be an Abelian torsion group. Then any variety in $\mathcal{C}(G)$ contains a character.*

Proof. Let V be any variety in $\mathcal{C}(G)$. Then V is equal to the annihilator of its annihilator, that is, there exists a subset Λ in the set $\mathcal{M}_c(G)$ of all finitely supported complex measures on G such that V is exactly the set of all functions in $\mathcal{C}(G)$ which are annihilated by all members of Λ:

$$V = V(\Lambda) = \{f \mid f \in C(G), \lambda(f) = 0 \text{ for all } \lambda \in \Lambda\}.$$

We show that for any nonempty finite subset Γ in Λ its annihilator $V(\Gamma)$ contains a character. Let F_Γ denote the subgroup generated by the supports of the measures belonging to Γ. Then F_Γ is a finitely generated torsion group. The measures belonging to Γ can be considered as measures on F_Γ and the annihilator of Γ in $C(F_\Gamma)$ will be denoted by $V(\Gamma)_{F_\Gamma}$. This is a variety in $\mathcal{C}(F_\Gamma)$. Indeed, if f belongs to V, then its restriction to F_Γ belongs to $V(\Gamma)_{F_\Gamma}$. If, in addition, we have $f(x_0) \neq 0$ and y_0 is in F_Γ, then the translate of f by $x_0 - y_0$ belongs to V, its restriction to F_Γ belongs to $V(\Gamma)_{F_\Gamma}$ and it takes the value $f(x_0) \neq 0$ at y_0. Hence $V(\Gamma)_{F_\Gamma}$ is a nonzero closed translation invariant subspace of $C(F_\Gamma)$, that is, a variety. As F_Γ is finitely generated, by Theorem 2.23 spectral synthesis holds for $C(F_\Gamma)$, and in particular, $V(\Gamma)_{F_\Gamma}$ contains nonzero exponential monomials. As F_Γ is a torsion group, any nonzero exponential monomial on F_Γ is a character. This means that $V(\Gamma)_{F_\Gamma}$ contains a character of F_Γ. By Theorem 3.1 any character of F_Γ can be extended to a character of G, and obviously any such extension belongs to $V(\Gamma)$.

Now we have proved that for any nonempty finite subset Γ in Λ the annihilator $V(\Gamma)$ contains a character. Let $char(V)$ denote the set of all characters contained in V. Obviously $char(V)$ is a compact subset of $\widehat{G}$, the dual of G, because $char(V)$ is closed and $\widehat{G}$ is compact. On the other hand, the system of nonempty compact sets $char(V(\Gamma))$, where Γ is a finite subset of Λ, has the finite intersection property:

$$char\big(V(\Gamma_1 \cup \Gamma_2)\big) \subseteq char\big(V(\Gamma_1)\big) \cap char(V(\Gamma_2)).$$

We infer that the intersection of this system is nonempty, and obviously

$$\emptyset \neq \bigcap_{\Gamma \subseteq \Lambda \text{ finite}} char\big(V(\Gamma)\big) \subseteq char(V).$$

This means that $char(V)$ is nonempty, and the theorem is proved.

3.2 Spectral analysis on Abelian groups

In the previous section we have seen that torsion groups behave nicely with respect to spectral analysis. Recently M. Laczkovich and G. Székelyhidi in [32] presented a complete characterization of Abelian groups having spectral analysis which clearly explains this nice behavior.

Let G be an Abelian group. We define the *torsion free rank* of G as the cardinality $r_0(G)$ of a maximal independent system of elements of infinite order. In other words, $r_0(G)$ is the maximal cardinality κ such that G contains the free Abelian group of rank κ as a subgroup. For instance, any commutative torsion group has torsion free rank 0. In [32] the authors prove the following theorem.

Theorem 3.4. *Spectral analysis holds on an Abelian group if and only if its torsion free rank is less than the continuum.*

This theorem is of basic importance: besides the complete characterization of Abelian groups having spectral analysis it clearly disproves the result of Elliot in [16] on spectral synthesis. Indeed, as spectral analysis is a consequence of spectral synthesis and there are Abelian groups without spectral analysis, namely those with torsion free rank larger than or equal to the continuum, hence spectral synthesis also fails to hold on these groups. Nevertheless, another earlier result in [75] already disproved Elliot's theorem by presenting a counterexample for an Abelian group without spectral synthesis — and even with torsion free rank less than the continuum. We will come back to this point in Section 3.5.

3.3 Spectral analysis on commutative semigroups

The basic ideas of spectral analysis and spectral synthesis can be formulated and investigated also on commutative *semigroups*. Let S be a locally compact commutative semigroup. The meaning of translation operators and translation invariant sets of functions is the same as in the case of groups and we use the same notation. It is easy to see that the relationship between varieties in $\mathcal{C}(S)$ and closed ideals in $\mathcal{M}_c(S)$ is the same: the annihilator of any variety in $\mathcal{C}(S)$ is a proper closed ideal in $\mathcal{M}_c(S)$, and conversely, the annihilator of any proper closed ideal in $\mathcal{M}_c(S)$ is a variety in $\mathcal{C}(S)$. Additive and exponential functions, as well as exponential monomials and exponential polynomials on S, have the same definition as in the group case. The basic question of spectral analysis is about the existence of an exponential in any variety, and similarly, the basic problem of spectral synthesis is if the exponential monomials in a given variety span a dense subspace. Although these problems are quite natural and the situation seems to be completely analogous to the group-case, however, there are some unexpected differences. To exhibit an interesting and important special case, we suppose that $S = \mathbb{N}^n$ with some positive integer n.

Let n be a fixed positive integer. For each $z = (z_1, z_2, \ldots, z_n)$ in $\mathbb{C}^n$ and for each multi-index $\alpha = (\alpha_1, \alpha_2, \ldots, \alpha_n)$ in $\mathbb{N}^n$ we will use the notation $z^\alpha = z_1^{\alpha_1} z_2^{\alpha_2} \ldots z_n^{\alpha_n}$ and $\alpha! = \alpha_1! \alpha_2! \ldots \alpha_n!$. If P is any complex polynomial in n variables, that is, any element of $\mathbb{C}[z_1, z_2, \ldots, z_n]$, the ring of all complex polynomials in n variables, then the notation for the differential operator $P(\partial) = P(\partial_1, \partial_2, \ldots, \partial_n)$ has the obvious meaning.

Using some simple ideas similar to those in the proofs of Theorem 6.10. and 6.11. in [70], p. 57., we have that any complex polynomial p on $\mathbb{N}^n$ is actually an ordinary polynomial in n complex variables and any exponential function m on $\mathbb{N}^n$ has the form

$$m(x_1, x_2, \ldots, x_n) = m_1(x_1) m_2(x_2) \ldots m_n(x_n)$$

for all $x_1, x_2, \ldots, x_n$ in $\mathbb{N}$ with some exponentials $m_i : \mathbb{N} \to \mathbb{C}$ of $\mathbb{N}$ $(i = 1, 2, \ldots, n)$. However, in contrast to the case of $\mathbb{Z}$, on $\mathbb{N}$ we have a special exponential m_0, which is 1 for $x = 0$ and is 0 for $x \neq 0$. We shall use the notation $m_0(x) = 0^x$ for this exponential, which is correct if we agree on $0^0 = 1$. This means that the exponentials of $\mathbb{N}^n$ have the form

$$m(x_1, x_2, \ldots, x_n) = \lambda_1^{x_1} \lambda_2^{x_2} \ldots \lambda_n^{x_n}$$

for each $x_1, x_2, \ldots, x_n$ in $\mathbb{N}$ with arbitrary complex numbers $\lambda_1, \lambda_2, \ldots, \lambda_n$. Hence the set of all exponentials of $\mathbb{N}^n$ can be identified with $\mathbb{C}^n$. We can use the notation λ^x for the product $\lambda_1^{x_1} \lambda_2^{x_2} \ldots \lambda_n^{x_n}$ if $\lambda = (\lambda_1, \lambda_2, \ldots, \lambda_n)$ and $x = (x_1, x_2, \ldots, x_n)$. For any finitely supported measure μ in $\mathcal{M}_c(\mathbb{N}^n)$ we will use its modified *Fourier–Laplace transform* which is now defined by

$$\widehat{\mu}(\lambda) = \int_{x \in \mathbb{N}^n} \lambda^x \, d\mu(x)$$

for all λ in $\mathbb{C}^n$. This is a polynomial in n complex variables. Obviously any polynomial of n complex variables is the Fourier–Laplace transform of some finitely supported measure on $\mathbb{N}^n$, hence the ring (actually algebra) of all Fourier–Laplace transforms of finitely supported measures on $\mathbb{N}^n$ can be identified with the ring $\mathbb{C}[z_1, z_2, \ldots, z_n]$. Basically, the Fourier–Laplace transformation $\mu \mapsto \hat{\mu}$ identifies $\mathcal{M}_c(\mathbb{N}^n)$ with the polynomial ring $\mathbb{C}[z_1, z_2, \ldots, z_n]$. The exponential corresponding to $\overline{\lambda}$ belongs to a variety if and only if λ is a common root of the polynomials corresponding to the annihilator ideal of the variety. By Hilbert's Nullstellensatz the polynomials in any proper ideal in $\mathbb{C}[z_1, z_2, \ldots, z_n]$ have a common root (see e.g. [79]), thus we have the following result.

Theorem 3.5. *Spectral analysis holds in $\mathbb{N}^n$.*

It turns out that spectral synthesis also holds for $\mathbb{N}^n$. To verify this statement one needs the famous Lasker–Noether Theorem on primary decomposition (see [79]), which states that in $\mathbb{C}[z_1, z_2, \ldots, z_n]$ each proper ideal is the intersection of finitely many primary ideals. Using this theorem a slight modification of the proof of Lefranc's theorem in [35] gives the following result.

Theorem 3.6. *Spectral synthesis holds in $\mathbb{N}^n$.*

3.4 Spectral synthesis and polynomial ideals

In this section we apply the results of the previous section to present a characterization theorem for polynomial ideals in several variables (see also [76]).

We have seen in the previous section that the ring of complex polynomials $\mathbb{C}[z_1, z_2, \ldots, z_n]$ can be identified with $\mathcal{M}_c(\mathbb{N}^n)$, the dual of $\mathcal{C}(\mathbb{N}^n)$, which is the space of all complex valued functions on $\mathbb{N}^n$ equipped with the topology of pointwise convergence. The weak*-topology on $\mathcal{M}_c(\mathbb{N}^n)$ is identical with the topology on $\mathbb{C}[z_1, z_2, \ldots, z_n]$ corresponding to componentwise convergence. Now we describe the identification between $\mathbb{C}[z_1, z_2, \ldots, z_n]$ and $\mathcal{M}_c(\mathbb{N}^n)$ in more detail.

Let p be a complex polynomial in $\mathbb{C}[z_1, z_2, \ldots, z_n]$. Writing z for the vector $(z_1, z_2, \ldots, z_n)$ the polynomial p can be written in the form

$$p(z) = \sum_{\alpha \in \mathbb{N}^n} \frac{1}{\alpha!} \partial^\alpha p(0) z^\alpha$$

for all z in $\mathbb{C}^n$, where $\alpha! = \alpha_1! \alpha_2! \ldots \alpha_n!$. Then the linear functional, or finitely supported measure μ_p, corresponding to p has its effect on a function f in $\mathcal{C}(\mathbb{N}^n)$ in the following way:

$$\langle \mu_p, f \rangle = \sum_{\alpha \in \mathbb{N}^n} \frac{1}{\alpha!} \partial^\alpha p(0) f(\alpha) \,.$$

Obviously, the convolution of μ_p and μ_r corresponds to $p \cdot r$. We observe that

$$\widehat{\mu_p}(\lambda) = \langle \mu_p(x), \lambda^x \rangle = \sum_{\alpha \in \mathbb{N}^n} \frac{1}{\alpha!} \partial^\alpha p(0) \lambda^x = p(\lambda) \,,$$

hence the Fourier–Laplace transform of μ_p can be identified with p. This means that we can write simply p for μ_p.

Now we take an exponential monomial $\varphi : x \mapsto p(x) \xi^x$ on $\mathbb{N}^n$ with some ξ in $\mathbb{C}^n$ and a polynomial p in $\mathbb{C}[z_1, z_2, \ldots, z_n]$. We have

$$p(\lambda) = \sum_{x \in \mathbb{N}^n} \frac{1}{x!} \partial^x p(0) \lambda^x \,,$$

hence

$$\partial^\alpha p(\lambda) = \sum_{x \in \mathbb{N}^n} \frac{1}{x!} \partial^x p(0) [x]^\alpha \lambda^{x-\alpha} \,,$$

where $[x]^\alpha$ denotes the product $[x_1]^{\alpha_1} [x_2]^{\alpha_2} \ldots [x_n]^{\alpha_n}$ with the notation $[x_i]^{\alpha_i} = x_i(x_i - 1) \ldots (x_i - \alpha_i + 1)$ for $i = 1, 2, \ldots, n$. It follows that

$$\lambda^\alpha \partial^\alpha p(\lambda) = \sum_{x \in \mathbb{N}^n} \frac{1}{x!} \partial^x p(0) [x]^\alpha \lambda^x \,.$$

The polynomial p has a unique representation in the form

$$p(x) = \sum_{\beta \in \mathbb{N}^n} c_\beta [x]^\beta \,,$$

which implies

$$\sum_{\beta \in \mathbb{N}^n} c_\beta \lambda^\beta \partial^\beta p(\lambda) = \sum_{x \in \mathbb{N}^n} \frac{1}{x!} \partial^x p(0) p(x) \lambda^x \,.$$

Suppose that μ_p annihilates φ, that is $\langle \mu_p, \varphi \rangle = 0$. This means

$$\langle \mu_p, \varphi \rangle = \sum_{x \in \mathbb{N}^n} \frac{1}{x!} \partial^x p(0) p(x) \xi^x = 0 \,,$$

or

$$\sum_{\beta \in \mathbb{N}^n} c_\beta \xi^\beta \partial^\beta p(\xi) = 0 \,. \tag{3.1}$$

In fact, this equation is necessary and sufficient for φ to be in the annihilator of the ideal generated by the polynomial p. Here ξ runs through the roots of the polynomial p. By Theorem 3.6 the linear hull of all exponential monomials of the form φ is dense in the annihilator of any proper ideal in $\mathbb{C}[z_1, z_2, \ldots, z_n]$. This means that a polynomial p belongs to a given proper ideal in $\mathbb{C}[z_1, z_2, \ldots, z_n]$ if and only if p satisfies a system of equations of the form (3.1), corresponding to the common roots of the polynomials in the given ideal and to different differential polynomials. We can formulate these results in the following theorem. We remark that for a proper ideal I in the polynomial ring $\mathbb{C}[z_1, z_2, \ldots, z_n]$ the set of all common roots of the polynomials in I is denoted by $Z(I)$. By Hilbert's Nullstellensatz (in other words: spectral analysis in $\mathbb{N}^n$) the set $Z(I)$ is nonempty.

Theorem 3.7. *(Ideal Theorem) Let I be a proper ideal in the polynomial ring $\mathbb{C}[z_1, z_2, \ldots, z_n]$. Then there exist nonempty sets of polynomials $\mathcal{P}_\xi$ for each ξ in $Z(I)$ such that a polynomial p belongs to I if and only if*

$$P(\partial)p\,(\xi) = 0 \tag{3.2}$$

holds for each ξ in $Z(I)$ and for each P in $\mathcal{P}_\xi$.

In the case $n = 1$ any proper ideal in $\mathbb{C}[z]$ is a principal ideal, hence $Z(I)$ is a nonempty finite set:

$$Z(I) = \{\xi_1, \xi_2, \ldots, \xi_k\},$$

where the complex numbers $\xi_1, \xi_2, \ldots, \xi_k$ are the different roots of the generating polynomial of I with positive multiplicities $m_1, m_2, \ldots, m_k$. In this case $\mathcal{P}_{\xi_j}$ can be taken as the set of polynomials $\{1, z, z^2, \ldots, z^{m_j - 1}\}$ for $j = 1, 2, \ldots, k$. The condition (3.2) means that a polynomial p belongs to I if and only if its derivatives $p^{(i)}$ for $i = 0, 1, \ldots, m_j - 1$ vanish at ξ_j for $j = 1, 2, \ldots, k$.

Now we describe the sets of polynomials $\mathcal{P}_\xi$ for a given proper ideal I in $\mathbb{C}[z_1, z_2, \ldots, z_n]$. We need the following simple result.

Theorem 3.8. *Let P, f, g be given polynomials in $\mathbb{C}[z_1, z_2, \ldots, z_n]$. Then we have*

$$P(\partial)(f \cdot g) = \sum_{\alpha \in \mathbb{N}^n} \frac{1}{\alpha!}\, \partial^\alpha f \cdot [(\partial^\alpha P)(\partial)]g\,. \tag{3.3}$$

Proof. The statement is obvious if P is a monomial of the form $P(z) = z^\beta$ by Leibniz's Rule:

$$\partial^\beta (f \cdot g) = \sum_{\alpha \in \mathbb{N}^n} \frac{\beta!}{\alpha!(\beta - \alpha)!}\, \partial^\alpha f \cdot \partial^{\beta - \alpha} g\,.$$

Hence the general statement follows.

Let ξ be a point in $Z(I)$ and let $\mathcal{P}_\xi(I)$ be the set of all polynomials P in $\mathbb{C}[z_1, z_2, \ldots, z_n]$ for which

$$P(\partial)f(\xi) = 0$$

holds for each f in I. Obviously $\mathcal{P}_\xi(I)$ is a linear space of polynomials, which is *closed under differentiation*: if P belongs to $\mathcal{P}_\xi(I)$ then $\partial^\alpha P$ belongs to $\mathcal{P}_\xi(I)$ for any multi-index α. On the other hand, using the Newton Interpolation Formula and the Taylor Formula we see that a linear space of polynomials is closed under differentiation if and only if it is translation invariant: derivatives are linear combinations of translates and translates are linear combinations of derivatives (see [61]). Hence for each ξ in $Z(I)$ the set $\mathcal{P}_\xi(I)$ is a translation invariant linear space of polynomials, which obviously contains the constant polynomials. The following theorem gives a complete description of $\mathcal{P}_\xi$.

Theorem 3.9. *Let I be a proper ideal in $\mathbb{C}[z_1, z_2, \ldots, z_n]$ and let ξ be a point in $Z(I)$. Then the polynomial P belongs to $\mathcal{P}_\xi(I)$ if and only if*

$$\sum_{\alpha \in \mathbb{N}^n} \frac{1}{\alpha!} \partial^\alpha f(\xi) \partial^\alpha P(z) = 0 \tag{3.4}$$

holds for each f in I and for all z in $\mathbb{C}^n$.

Proof. Obviously we may suppose that $I \neq \{0\}$. First suppose that P satisfies (3.4) for each f in I and for each z in $\mathbb{C}^n$. For any multi-index α and for any z in $\mathbb{C}^n$ we let $q_\alpha(z) = (z - \xi)^\alpha$. Then it follows that

$$[P(\partial)q_\alpha](\xi) = \partial^\alpha P(0), \tag{3.5}$$

hence

$$P(\partial)f(\xi) = P(\partial)\Big[\sum_{\alpha \in \mathbb{N}^n} \frac{1}{\alpha!} \partial^\alpha f(\xi)q_\alpha\Big](\xi)$$

$$= \sum_{\alpha \in \mathbb{N}^n} \frac{1}{\alpha!} \partial^\alpha f(\xi)[P(\partial)q_\alpha](\xi) = \sum_{\alpha \in \mathbb{N}^n} \frac{1}{\alpha!} \partial^\alpha f(\xi) \partial^\alpha P(0) = 0,$$

for any f in I by (3.4). This means that P is in $\mathcal{P}_\xi(I)$.

Conversely, suppose that P is in $\mathcal{P}_\xi(I)$. Then we have as above

$$0 = P(\partial)f(\xi) = P(\partial)\Big[\sum_{\alpha \in \mathbb{N}^n} \frac{1}{\alpha!} \partial^\alpha f(\xi)q_\alpha\Big](\xi)$$

$$= \sum_{\alpha \in \mathbb{N}^n} \frac{1}{\alpha!} \partial^\alpha f(\xi)[P(\partial)q_\alpha](\xi) = \sum_{\alpha \in \mathbb{N}^n} \frac{1}{\alpha!} \partial^\alpha f(\xi) \partial^\alpha P(0).$$

As $\mathcal{P}_\xi(I)$ is translation invariant, this latter equation holds for any translate of P. Replacing P by $w \mapsto P(w + z)$ our statement follows.

For any translation invariant linear space L of polynomials in the ring $\mathbb{C}[z_1, z_2, \ldots, z_n]$ and for each ξ in $\mathbb{C}^n$ let $\mathcal{I}_\xi(L)$ denote the set of all polynomials f in $\mathbb{C}[z_1, z_2, \ldots, z_n]$ for which (3.4) holds for all z in $\mathbb{C}^n$. We have the following simple result.

Theorem 3.10. *For any nonzero translation invariant linear subspace L in $\mathbb{C}[z_1, z_2, \ldots, z_n]$ and for any ξ in $\mathbb{C}^n$ the set $\mathcal{I}_\xi(L)$ is a proper ideal in the ring $\mathbb{C}[z_1, z_2, \ldots, z_n]$.*

Proof. Let L be a nonzero translation invariant linear subspace in the ring $\mathbb{C}[z_1, z_2, \ldots, z_n]$ and let z be any element in $\mathbb{C}^n$. Then L is closed under differentiation, hence it contains the constant polynomials. Let f be an element of $\mathcal{I}_\xi(L)$ and let p be any polynomial in $\mathbb{C}[z_1, z_2, \ldots, z_n]$. Then by Theorem 3.8 we have for each P in L,

$$P(\partial)(f \cdot p)(\xi) = \sum_{\alpha \in \mathbb{N}^n} \frac{1}{\alpha!} \partial^\alpha p(\xi)[\partial^\alpha P(\partial)f](\xi).$$

As $\partial^\alpha P$ belongs to L for any multi-index α, it follows that $[\partial^\alpha P(\partial)f](\xi) = 0$ for any α, hence $f \cdot p$ belongs to $\mathcal{I}_\xi(L)$. This means that $\mathcal{I}_\xi(L)$ is an ideal, and as the constants belong to L, each element of $\mathcal{I}_\xi(L)$ vanishes at ξ. Thus $\mathcal{I}_\xi(L)$ is proper.

By Theorem 3.7 and by this latter theorem we have that any proper ideal in the ring $\mathbb{C}[z_1, z_2, \ldots, z_n]$ is the intersection of ideals of the form $\mathcal{I}_\xi(L)$ with some nonzero translation invariant linear spaces of polynomials L and some elements ξ in $\mathbb{C}^n$. By the Hilbert Basis Theorem (see e.g. [79] Vol. I., p. 200.) any proper ideal in $\mathbb{C}[z_1, z_2, \ldots, z_n]$ is finitely generated. Suppose that I is generated by the polynomials $f_1, f_2, \ldots, f_k$, where k is a positive integer. Then $Z(I)$ is the set of all common zeros of the polynomials $f_1, f_2, \ldots, f_k$. For any ξ in $Z(I)$ we consider the the system of partial differential equations

$$\sum_{\alpha \in \mathbb{N}^n} \frac{1}{\alpha!} \partial^\alpha f_j(\xi) \partial^\alpha P(z) = 0 \tag{3.6}$$

for any z in $\mathbb{C}^n$ and for $j = 1, 2, \ldots, k$. The set of all polynomial solutions of this system is $\mathcal{P}_\xi(I)$. In some sense $\mathcal{P}_\xi(I)$ can be considered as the *multiplicity* of the ideal I at the common zero ξ.

In the case $n = 1$ let the proper ideal I in $\mathbb{C}[z]$ be generated by the polynomial f of degree N and let ξ be any root of f with multiplicity $m \leq N$. Then the system (3.6) has the form

$$\sum_{i=m}^{N} \frac{1}{i!} f^{(i)}(\xi) P^{(i)}(z) = 0$$

for all z in $\mathbb{C}$. As $f^{(m)}(\xi) \neq 0$, hence (3.4) is equivalent to the ordinary differential equation $P^{(m)}(z) = 0$ and the set of all polynomial solutions is the set of all polynomials of degree at most $m - 1$.

It seems to be possible to characterize different types of ideals in the polynomial ring $\mathbb{C}[z_1, z_2, \ldots, z_n]$, like prime ideals, primary ideals, radical ideals, etc., in terms of the *multiplicity spaces* $\mathcal{P}_\xi(I)$. For instance, I is a maximal ideal if and only if $Z(I) = \{\xi\}$ is a singleton and $\mathcal{P}_\xi(I)$ is one dimensional. In particular, the Ideal Membership Problem can be solved in the following form: the proper ideal I is contained in the proper ideal J if and only if $Z(J)$ is a subset of $Z(I)$ and $\mathcal{P}_\xi(J)$ is a subset of $\mathcal{P}_\xi(I)$ for each ξ in $Z(J)$.

We remark that characterization of polynomial ideals in several variables is the content of the Ehrenpreis–Palamodov theorem (see [59], Theorem 10.12., p. 141.). One of its consequences is the following theorem (see [59], Theorem 10.13., p. 142.).

Theorem 3.11. *Given any primary ideal I in the ring of complex polynomials in n variables, there exist differential operators with polynomial coefficients*

$$A_i(x, \partial) = \sum_j p_j^i(x_1, x_2, \ldots, x_n) \partial_1^{j_1} \partial_2^{j_2} \ldots \partial_n^{j_n}$$

for $i = 1, 2, \ldots, r$ with the following property: a polynomial f lies in the ideal I if and only if the result of applying $A_i(x, \partial)$ to f vanishes on the variety of I for $i = 1, 2, \ldots, r$.

The *variety of a polynomial ideal* is the set of all common zeros of the polynomials in the ideal. The differential operators $A_1(x, \partial), A_2(x, \partial), \ldots, A_r(x, \partial)$ are called *Noetherian operators* for the primary ideal I. An algorithm for computing Noetherian operators for a given primary ideal is given in [44]. Hence our approach presented above can be considered as an alternative way to find Noetherian operators.

3.5 The failure of spectral synthesis on some types of discrete Abelian groups

Let G be an Abelian group. The function $B : G \times G \to \mathbb{C}$ is called *bi-additive* if the functions $x \mapsto B(x, y)$ and $x \mapsto B(y, x)$ are additive for each fixed y in G. It is called *symmetric* if $B(x, y) = B(y, x)$ for all x, y in G. We shall use the *difference operators* Δ_y for each y in G in the usual way: given a complex valued function f on G we let

$$\Delta_y f(x) = f(x + y) - f(x)$$

for each x in G. Then $\Delta_y f$ is a complex valued function on G. Symbolically we can write

$$\Delta_y = \tau_y - \tau_0,$$

where 0 is the zero element of G. Iterates of Δ_y have the obvious meaning. For instance,

$$\Delta_y^2 f(x) = (\Delta_y \circ \Delta_y) f(x) = f(x + 2y) - 2f(x + y) + f(x),$$

or

$$\Delta_y^3 f(x) = (\Delta_y \circ \Delta_y^2) f(x) = f(x + 3y) - 3f(x + 2y) + 3f(x + y) - f(x)$$

holds for any complex valued function f on G and for each x, y in G.

First we prove the following theorem.

Theorem 3.12. *Let G be an Abelian group. If there exists a symmetric bi-additive function $B : G \times G \to \mathbb{C}$ such that the variety V generated by the function $x \mapsto B(x, x)$ is of infinite dimension, then spectral synthesis fails to hold in V.*

Proof. Let $f(x) = B(x, x)$ for all x in G. By the equation

$$f(x + y) = B(x + y, x + y) = B(x, x) + 2B(x, y) + B(y, y) \qquad (3.7)$$

we see that the translation invariant subspace, generated by f, as a linear space is generated by the functions 1, f and all the additive functions of the form $x \mapsto B(x, y)$, where y runs through G. Hence our assumption on B is equivalent to the condition that there are infinitely many functions of the form $x \mapsto B(x, y)$ with y in G, which are linearly independent. This also implies that there is no positive integer n such that B can be represented in the form

$$B(x, y) = \sum_{k=1}^{n} a_k(x) b_k(y),$$

where $a_k, b_k : G \to \mathbb{C}$ are additive functions $(k = 1, 2, \ldots, n)$. Indeed, the existence of a representation of this form would mean that the number of linearly independent additive functions of the form $x \mapsto B(x, y)$ is at most n.

It is clear that any translate of f, hence any function g in V, satisfies

$$\Delta_y^3 g(x) = 0 \qquad (3.8)$$

for all x, y in G: this can be checked for f. Hence any exponential m in V satisfies the same equation, which implies

$$m(x)\big(m(y) - 1\big)^3 = 0$$

for all x, y in G, and this means that m is identically 1. It follows that any exponential monomial in V is a polynomial, and by (3.8) it is of degree at most 2. On the other hand, suppose that p is a polynomial of degree 2 in V of the form

$$p(x) = \sum_{k=1}^{n} \sum_{l=1}^{m} c_{kl} \, a_k(x) b_l(x) + c(x) + d = p_2(x) + c(x) + d$$

with some positive integers n, m, additive functions $a_k, b_l, c \, : \, G \to \mathbb{C}$ and constant d, where p_2 is not identically zero. By assumption, p is the pointwise limit of a net formed by linear combinations of translates of f, that means, by functions of the form (3.7). Linear combinations of functions of the form (3.7) have the form

$$\varphi(x) = cB(x, x) + A(x) + D \,,$$

with some additive function $A \, : \, G \to \mathbb{C}$, and constants c, D. Any net formed by these functions has the form

$$\varphi_\gamma(x) = c_\gamma B(x, x) + A_\gamma(x) + D_\gamma \,.$$

By pointwise convergence

$$\lim_\gamma \frac{1}{2} \Delta_y^2 \varphi_\gamma(x) = \frac{1}{2} \Delta_y^2 p(x) = p_2(y)$$

follows for all x, y in G. On the other hand,

$$\lim_\gamma \frac{1}{2} \Delta_y^2 \varphi_\gamma(x) = B(y, y) \lim_\gamma c_\gamma \,,$$

holds for all x, y in G, hence the limit $\lim_\gamma c_\gamma = c$ exists and is different from zero, which gives $B(x, x) = \frac{1}{c} p_2(x)$ for all x in G, and this is impossible.

We infer that any exponential monomial φ in V is actually a polynomial of degree at most 1, which satisfies

$$\Delta_y^2 \varphi(x) = 0 \tag{3.9}$$

for all x, y in G, hence any function in the closed linear hull of the exponential monomials in V satisfies this equation. However f does not satisfy (3.9), hence the linear hull of the exponential monomials in V is not dense in V.

Now we are in the position to present an example for an Abelian group, where a bi-additive function is available, subjected to the above conditions.

Theorem 3.13. *Spectral synthesis fails to hold for the additive group of real numbers.*

Proof. We show that if $G = \mathbb{R}$ is the additive group of the real numbers, then there exists a symmetric bi-additive function $B : \mathbb{R} \times \mathbb{R} \to \mathbb{C}$ with the property that there are infinitely many linearly independent functions of the form $x \mapsto B(x, y)$, where y is in $\mathbb{R}$. Let H denote a basis of the linear space $\mathbb{R}$ over the rationals, which is sometimes called a Hamel basis. For any ξ in H let p_ξ denote the projection of the linear space $\mathbb{R}$ over $\mathbb{Q}$ onto the one-dimensional subspace generated by ξ. This means that for any real number x the rational number $p_\xi(x)$ is the coefficient of x in the unique representation with respect to the basis H. It is clear that the functions p_ξ are additive and linearly independent for different choices of ξ in H. Let

$$B(x, y) = \sum_{\xi \in H} p_\xi(x) p_\xi(y)$$

for each x, y in $\mathbb{R}$. The sum is finite for any fixed x, y, and obviously B is symmetric and bi-additive. On the other hand, if χ is any element of H, then we have

$$B(x, \chi) = \sum_{\xi \in H} p_\xi(x) p_\xi(\chi) = p_\chi(x),$$

hence the functions $x \mapsto B(x, \chi)$ are linearly independent for different elements χ in H. Our theorem is proved.

From the above proof it is clear that the same construction works on any Abelian group which has a subgroup isomorphic to the (non-complete) direct sum of infinitely many copies of the additive group of integers. Spectral synthesis fails to hold on any of those Abelian groups as the following theorem states (see [75]).

Theorem 3.14. *Spectral synthesis fails to hold on any Abelian group with infinite torsion free rank.*

This means that a necessary condition for spectral synthesis is that the torsion free rank is finite. By Lefranc's result in [35] a sufficient condition is that the group is finitely generated. Based on these facts we can formulate two quite reasonable conjectures.

Conjecture 1. Spectral synthesis holds on an Abelian group if and only if it is finitely generated.

Conjecture 2. Spectral synthesis holds on an Abelian group if and only if its torsion free rank is finite.

In the following section we will disprove Conjecture 1: there are Abelian groups with spectral synthesis without a finite generating set.

3.6 Spectral synthesis on Abelian torsion groups

In this section we prove that spectral synthesis holds on commutative torsion groups. This result clearly disproves Conjecture 1, formulated in the previous section, because it shows that there are Abelian groups without a finite generating set such that spectral synthesis still holds on them.

First of all we remark that the set of all finitely supported complex measures on an Abelian group can be identified with the *finite group algebra* $\mathbb{C}G$ of the group. We have the following theorem.

Theorem 3.15. *Given an Abelian torsion group, then all characters in a nonzero variety generate a dense subspace in this variety if and only if its annihilator ideal in the finite group algebra of the group is the intersection of all maximal ideals including it.*

Proof. First we prove the following statement: if G is an Abelian torsion group, then each maximal ideal M of the finite group algebra $\mathbb{C}G$ has the following form: there exists a character γ_M of G such that the measure x belongs to M if and only if $\langle x, \gamma_M \rangle = 0$. We remark that the converse statement is obvious: if M is an ideal of this form, then its annihilator is the one-dimensional linear subspace generated by γ_M, hence M clearly cannot be included in any proper closed ideal.

Suppose now that M is a maximal ideal in $\mathbb{C}G$. Then $\mathbb{C}G/M$ is a field, which is an extension of the complex field, as the natural homomorphism Φ of $\mathbb{C}G$ onto $\mathbb{C}G/M$ restricted to the constant multiples of the identity in $\mathbb{C}G$ sets up a field isomorphism onto a subfield of $\mathbb{C}G/M$, which is isomorphic to $\mathbb{C}$. On the other hand, let g be any element of G and n a positive integer with the property that $g^n = 1$. Then

$$\delta_g^{*n} = \delta_{g^n} = \delta_1 \,,$$

and hence

$$\Phi(\delta_g)^n = \Phi(\delta_g^{*n}) = \Phi(\delta_1) = 1 \,,$$

consequently $\Phi(\delta_g)$ is a complex n-th root of unity for any g in G. In particular, $\Phi(\delta_g)$ is a complex number, hence the function $\gamma : G \to \mathbb{C}$ defined by

$$\gamma(g) = \Phi(\delta_g)$$

for each g in G is a homomorphism of G into the complex unit circle, that is, a character of G. Clearly x belongs to M if and only if $\Phi(x) = 0$, which means that $\langle x, \overline{\gamma} \rangle = 0$. Choosing $\gamma_M = \overline{\gamma}$ our first statement is proved.

Suppose now that all characters in the nonzero variety V generate a dense subspace in V and let I denote the annihilator of V, which is a proper closed ideal in $\mathbb{C}G$. If x is any element belonging to each maximal ideal including I, then the above considerations show that x annihilates all characters which are included in V, and by our assumption, it follows that x annihilates V. Hence x belongs to I. Conversely, suppose that I, the annihilator of V, is the intersection of all maximal ideals including I. Suppose that the subvariety generated by all characters in V is smaller than V. Then by the Hahn–Banach Theorem there exists a linear functional x in $\mathbb{C}G$ which annihilates all characters in V, but it does not belong to I. Annihilating all characters in V means that x belongs to all maximal ideals including I, hence, by our assumption it must belong to I, which is a contradiction and our theorem is proved.

We shall make use of the following result which is a consequence of Theorem 2.23 but here we give an independent proof.

Theorem 3.16. *Spectral synthesis holds on any finite Abelian group.*

Proof. If G is a finite Abelian group, then $\mathbb{C}^G$ is a finite dimensional linear space in which every linear subspace has a direct complement. It is easy to see that the direct complement of any variety is a variety, too. Indeed, if V is a translation invariant subspace of $\mathbb{C}^G$, then any element f in $\mathbb{C}^G$ can be represented uniquely in the form $f = v + w$, where v is in V and w is in its direct complement W. Let u be any element in W and let h be any element of G. Then $\tau_h u$ has a representation $\tau_h u = a + b$ with a in V and b in W. This means $u = \tau_{-h}a + \tau_{-h}b$, and here $\tau_{-h}a$ belongs to V by the translation invariance of this space. As u is in W and the above representation is unique, we have $\tau_{-h}a = 0$ and $a = 0$, which implies that $\tau_h u$ belongs to W, hence W is a variety.

Suppose now that the closed linear hull V_0 of all characters in a nonzero variety V on G is different from V. The intersection of the direct complement of V_0 in $\mathbb{C}^G$ with V is a nonzero subvariety in V. By Theorem 3.3 about spectral analysis on Abelian torsion groups, this variety contains a character of G, which belongs to V and this is a contradiction, as it does not belong to V_0.

Theorem 3.17. *Spectral synthesis holds on any Abelian torsion group.*

Proof. By the above considerations it is enough to prove that in $\mathbb{C}G$ any proper closed ideal is the intersection of all maximal ideals in which it is included. For any ideal I in $\mathbb{C}G$ and for any finite subset K in G, let I_K denote the set of all measures in I with support in the subgroup G_K of G generated by the set K. Clearly $I = \bigcup_{K \in \mathcal{K}} I_K$, where $\mathcal{K}$ denotes the set of all finite subsets of G. For any function f in $\mathbb{C}^G$ the restriction of f to the subgroup G_K generated by K will be denoted by $f|_K$. If x is any measure in I_K, then $x|_K$ is a measure in $\mathbb{C}G_K$, hence I_K can be considered as a subset of $\mathbb{C}G_K$ and it is obviously a closed linear subspace. If y is any measure in

$\mathbb{C}G_K$, then we can extend it to a measure $\tilde{y}$ in $\mathbb{C}G$ by assigning zero to any g in G not belonging to G_K. It follows that $x * \tilde{y}$ belongs to I and has its support in G_K, hence it belongs to I_K, which means that I_K is a closed ideal in $\mathbb{C}G_K$. Clearly I_K is proper if so is I. As G_K is a finite Abelian group, spectral synthesis holds on it by our previous theorem.

Suppose that I is a proper closed ideal in $\mathbb{C}G$, which is not the intersection of all maximal ideals including it. This means that some x in the intersection of all maximal ideals containing I does not belong to I, hence it does not belong to any I_K, where K is a finite subset of G. Let J denote the support of this x. As spectral synthesis holds on G_K for any finite subset K of G, this means that all characters in $I_K^\perp$ generate a dense subspace. As x does not belong to I_K for any finite K, it follows that for any finite subset K of G including J there exists a character γ_K of G_K such that $\langle x, \gamma_K \rangle \neq 0$, but $\langle y, \gamma_K \rangle = 0$ for each y in I with support in K. As any character of G_K can be extended to a character of G, by fixing such an extension for each K we will suppose that γ_K actually denotes this extension. Now we consider the net $(\gamma_K)_{K \in \mathcal{K}_0}$ along the directed set $\mathcal{K}_0$ of finite subsets of G including J. This net lies in the character group of G, which is a compact topological space, hence the net has a convergent sub-net $(\gamma_K)_{K \in \mathcal{K}_1}$ converging pointwise to a character γ_0 of G. Here $\mathcal{K}_1$ is a co-final subset of $\mathcal{K}_0$, that is, for any finite subset K of G in $\mathcal{K}_0$ there exists a finite subset K_1 in $\mathcal{K}_1$ with $K \subseteq K_1$. Convergence of the sub-net $(\gamma_K)_{K \in \mathcal{K}_1}$ to γ_0 implies pointwise convergence of the corresponding restrictions on any finitely generated subgroup, that is, if L is any finite subset of G, then the net of the restrictions $(\gamma_K|_L)_{K \in \mathcal{K}_1}$ converges to $\gamma_0|_L$ pointwise on G_L. The restrictions $\gamma_K|_L$ for K in $\mathcal{K}_1$ are characters of G_K, that is, elements of the finite character group of G_K. By pointwise convergence this means that they must satisfy the following property $\mathcal{P}$: for any finite subset L of G there exists a finite subset K_0 of G in $\mathcal{K}_1$ such that for each finite subset K in $\mathcal{K}_1$ including K_0 we have that $\gamma_K|_L = \gamma_0|_L$. First we apply this property for the finite set J, the support of x. Let K_0 be a finite subset of G including J such that for each finite subset K in $\mathcal{K}_1$ including K_0 we have that $\gamma_K|_J = \gamma_0|_J$. Then it follows that

$$\langle x, \gamma_0 \rangle = \langle x|_J, \gamma_0|_J \rangle = \langle x|_J, \gamma_K|_J \rangle = \langle x|_K, \gamma_K|_J \rangle = \langle x, \gamma_K \rangle \neq 0 .$$

Let now y be any element of I with support L, that is, y is in I_L. We apply property $\mathcal{P}$ for the finite subset L: let K_0 be a finite subset of G including J such that for each finite subset K in $\mathcal{K}_1$ including K_0 we have that $\gamma_K|_L = \gamma_0|_L$. We take any finite set K in G including L and K_0, then the support of y is in K, hence $\langle y, \gamma_K \rangle = 0$ and

$$\langle y, \gamma_0 \rangle = \langle y|_L, \gamma_0|_L \rangle = \langle y|_L, \gamma_K|_L \rangle = \langle y, \gamma_K \rangle = 0 .$$

We have proved that the maximal ideal in $\mathbb{C}G$ corresponding to the character γ_0 does not contain x, however it contains I, which is a contradiction and our theorem is proved.

Actually, the property that each proper ideal of the commutative group algebra is the intersection of all maximal ideals including it characterizes Abelian torsion groups as the following theorem states (see [3]).

Theorem 3.18. *Let G be an Abelian group. Then G is a torsion group if and only if each proper ideal of the complex group algebra $\mathbb{C}G$ is an intersection of maximal ideals.*

We can reformulate Theorem 3.17 in the following way, obtaining an analogue of Hilbert's Nullstellensatz (see [79]).

Theorem 3.19. *(Nullstellensatz) Suppose, that a nonempty set of trigonometric polynomials on a 0-dimensional compact Abelian group is given, and another trigonometric polynomial is zero on all the common roots of the trigonometric polynomials belonging to the given set. Then this trigonometric polynomial is included in the ideal generated by the given set of trigonometric polynomials.*

Proof. By duality theory any 0-dimensional compact Abelian group is the dual of an Abelian torsion group G (see e.g. [24]). Any trigonometric polynomial on $\widehat{G}$ is a finite linear combination of characters of $\widehat{G}$, that is, the Fourier transform of a finitely supported measure on G. Hence the statement of the present theorem can be reformulated in the following way: if a finitely supported measure on G annihilates all characters, which are annihilated by a given nonempty set of finitely supported measures, then it belongs to the ideal in $\mathbb{C}G$ generated by the given set. But this is exactly spectral synthesis on G and our theorem is proved.

There is another way to formulate our theorem, which may enlighten its relation to the Beurling problem mentioned in Section 2.3 (see [5]). Namely, our spectral synthesis theorem implies the following result.

Theorem 3.20. *Let G be an Abelian torsion group. Given a complex-valued function f on G the finitely supported measure x satisfies $x * f = 0$ if and only if $\hat{x}$ vanishes on the spectral set of f.*

Proof. Let the finitely supported measure x satisfy $x * f = 0$ for some function $f : G \to \mathbb{C}$. Then x belongs to the annihilator of the variety generated by f, hence x annihilates all characters belonging to this variety. In other words, $\hat{x}$ vanishes on the spectral set of f. Conversely, suppose that $\hat{x}$ vanishes on the characters belonging to the variety generated by f. By spectral synthesis these characters span a dense linear subspace of the variety, hence x annihilates the variety and $x * f = 0$.

We also have the following analogue of the Primary Ideal Theorem.

Theorem 3.21. *Let G be an Abelian torsion group. If the spectral set of a complex-valued function consists of a single point, then the function is a constant multiple of a character. If the spectral set of a complex-valued function is finite, then the function is a trigonometric polynomial.*

3.7 Polynomial functions and spectral synthesis

In this section we study the connection of spectral synthesis and polynomial functions (see also [75]). In particular, we shall present an equivalent formulation of Conjecture 2 given in Section 3.5. Polynomial functions naturally appear in the study of spectral analysis and spectral synthesis problems on Abelian groups. In Section 3.5 we have seen that spectral synthesis fails to hold on any Abelian group on which there exists a certain generalized polynomial function. Here we present a characterization of Abelian groups with finite torsion free rank in terms of polynomial functions. It turns out that the torsion free rank of an Abelian group is finite if and only if each complex generalized polynomial on the group is actually a polynomial. Hence the violation of spectral synthesis is basically due to the existence of "strange" polynomials. For instance, on commutative torsion groups we have an extreme situation: any complex polynomial is constant, and this is in complete accordance with the results Section 3.1: spectral synthesis holds on any commutative torsion group. However, the question about the sufficiency of the non-existence of "strange" polynomials for the presence of spectral synthesis remains open.

Let G be an Abelian group. In Section 3.5 we defined the difference operators Δ_y and $\Delta_{y_1,y_2,\ldots,y_n}$ for the product $\Delta_{y_1} \circ \Delta_{y_2} \circ \cdots \circ \Delta_{y_n}$. In particular, if $y_1 = y_2 = \cdots = y_n$, then we write Δ_y^n for $\Delta_{y_1,y_2,\ldots,y_n}$. More explicitly, for any function $f : G \to \mathbb{C}$ and for any x, y in G we have

$$\Delta_y^n f(x) = \sum_{k=0}^n \binom{n}{k} (-1)^{n-k} f(x + ky). \tag{3.10}$$

The functional equation

$$\Delta_{y_1,y_2,\ldots,y_{n+1}} f(x) = 0 \tag{3.11}$$

is called *Fréchet's equation*, the functional equation

$$\Delta_y^{n+1} f(x) = 0 \tag{3.12}$$

is called the *polynomial equation*, and the functional equation

$$\Delta_y^n f(x) = n! f(y) \tag{3.13}$$

is called the *monomial equation*. We suppose here that $f : G \to \mathbb{C}$ is the unknown function and the equations hold for all $x, y, y_1, y_2, \ldots, y_{n+1}$, respectively. Solutions of the polynomial equation (3.12) are called *complex generalized polynomials of degree at most n* and nonzero solutions of the monomial equation (3.13) are called *complex generalized monomials of degree n*. Sometimes the zero function is also considered a generalized monomial without degree. It is clear that (3.11) implies (3.12). It is less obvious that (3.12) implies (3.11), that is, Fréchet's equation and the polynomial equation (with the same n) are equivalent. This is a consequence of a theorem of Djokovič (see [12]). It is also known that any solution f of (3.12) has a unique representation in the form

$$f(x) = \sum_{j=0}^{n} a_j(x) \tag{3.14}$$

for all x in G, where $a_j : G \to \mathbb{C}$ is a solution of (3.13) with j in place of n. In other words, any nonzero complex generalized polynomial of degree at most n has a unique representation as a sum of nonzero complex generalized monomials of degree not higher than n. In this representation a_j is called the *homogeneous term of degree j of f*. The nonzero homogeneous term of the highest degree of f is called the *leading term* of f and its degree is called the *degree of f*. The zero function has no degree. If f is a nonzero complex generalized polynomial of degree n, then the function $x \mapsto \Delta_y^n f(x)$ is constant for any fixed y in G and the leading term of f is the complex generalized monomial

$$a_n(x) = \frac{1}{n!} \, \Delta_x^n f(y) \tag{3.15}$$

for any x, y in G. These results also follow from the theorems in [12] (see also [66]). Another consequence of these results is that if a complex generalized polynomial is identically zero, then its homogeneous term of each degree must be zero, and if two nonzero complex generalized polynomials are identical, then they must have the same degree and the corresponding homogeneous terms of each degree must be identical. Roughly speaking, a kind of "comparing the coefficients"-like method works for complex generalized polynomials.

Let n be a positive integer. If G is any set and a function $F : G^n \to \mathbb{C}$ is given, then the function $x \mapsto F(x, x, \ldots, x)$ is called the *diagonalization of F*. The function $F : G^n \to \mathbb{C}$ is called *symmetric* if

$$F(x_{\sigma(1)}, x_{\sigma(2)}, \ldots, x_{\sigma(n)}) = F(x_1, x_2, \ldots, x_n)$$

holds for any $x_1, x_2, \ldots, x_n$ in G and for any permutation σ of the set $\{1, 2, \ldots, n\}$.

If G is an Abelian group, then the function $F : G^n \to \mathbb{C}$ is *n-additive* if the function $t \mapsto F(x_1, \ldots, x_{i-1}, t, x_{i+1}, \ldots, x_n)$ is a homomorphism of G into the additive group of complex numbers for any $i = 1, 2, \ldots, n$ and for any $x_1, \ldots, x_{i-1}, x_{i+1}, \ldots, x_n$. We call 1-additive functions simply *additive*. Sometimes this terminology is extended for $n = 0$ by considering any constant function to be 0-additive. If σ is any permutation of the set $\{1, 2, \ldots, n\}$, then the function $\sigma F : G^n \to \mathbb{C}$ defined by

$$\sigma F(x_1, x_2, \ldots, x_n) = \frac{1}{n!} \sum_{\sigma} F(x_{\sigma(1)}, x_{\sigma(2)}, \ldots, x_{\sigma(n)})$$

(the summation extends for all permutations σ of the set $\{1, 2, \ldots, n\}$) is obviously symmetric and has the same diagonalization as F. Moreover, if F is n-additive then σF is n-additive, too. In the case $n = 2$ instead of 2-additive we use the term *bi-additive* as in Section 3.5. For instance, a special type of complex bi-additive functions can be obtained in the following way: let n be a positive integer, $a_1, a_2, \ldots, a_n$ complex additive functions and let b_{ij} be complex numbers for $i, j = 1, 2, \ldots, n$. Then the function $B : G \times G \to \mathbb{C}$ defined by

$$B(x, y) = \sum_{i=1}^{n} \sum_{j=1}^{n} b_{ij} a_i(x) a_j(y)$$

for each x, y in G is a complex bi-additive function, which is a *bilinear function of complex additive functions*. Nevertheless, it is not true that any complex bi-additive function is a bilinear function of complex additive functions.

In [12] (see also [66]) it is proved that if n is a positive integer and G is an Abelian group, then the diagonalization of any nonzero n-additive symmetric function $F : G^n \to \mathbb{C}$ is a complex generalized monomial of degree n. Taking σF instead of F we see that this holds for the diagonalization of any n-additive function. Further from the results of [12] it follows that the converse of this statement is also true: any complex generalized monomial of degree n is the diagonalization of some n-additive symmetric function.

In the class of complex generalized polynomials we have the special subclass of polynomials. Polynomials are the elements of the complex algebra generated by all complex homomorphisms.

Let G be an Abelian group and consider $\mathbb{C}^G$, the set of all complex-valued functions on G. It is clear that the variety generated by a complex generalized polynomial, resp. a polynomial consists of complex generalized polynomials, resp. polynomials. The following theorem shows that polynomials can be characterized as complex generalized polynomials generating finite dimensional varieties.

Theorem 3.22. *A complex generalized polynomial on an Abelian group is a polynomial if and only if it generates a finite dimensional variety.*

Proof. First we show that any complex polynomial generates a finite dimensional variety. Let n be a positive integer, P a complex polynomial in n variables, $a_1, a_2, \ldots, a_n$ complex additive functions on the Abelian group G and consider the complex polynomial defined by

$$f(x) = P\big(a_1(x), a_2(x), \ldots, a_n(x)\big)$$

for all x in G. By the Taylor Formula we have, for each x, y in G,

$$f(x + y) =$$

$$\sum_{\alpha_1 + \cdots + \alpha_n \leq N} \frac{1}{\alpha_1! \ldots \alpha_n!}\, \partial_1^{\alpha_1} \ldots \partial_n^{\alpha_n} P\big(a_1(x), \ldots, a_n(x)\big) a_1(y)^{\alpha_1} \ldots a_n(y)^{\alpha_n}.$$

Here $\alpha_1, \alpha_2, \ldots, \alpha_n$ are nonnegative integers and N is the degree of the polynomial P. This equation shows that the linear space generated by all translates of f is of finite dimension. As finite dimensional subspaces of locally convex topological vector spaces are closed, our statement is proved.

For the converse suppose that the complex generalized polynomial f on G generates a finite dimensional variety. Then there is a nonnegative integer n and there are functions $g_i, h_i : G \to \mathbb{C}$ $(i = 1, 2, \ldots, n)$ such that the functional equation

$$f(x + y) = \sum_{i=1}^n g_i(x) h_i(y)$$

holds for any x, y in G. By Theorem 5.2.1. in [66] it follows that f is a normal exponential polynomial on G, which means that it has the form

$$f(x) = \sum_{k=1}^l P_k(x) m_k(x),$$

where $P_k : G \to \mathbb{C}$ is a polynomial and $m_k : G \to \mathbb{C}$ is an exponential, that is, a homomorphism of G into the multiplicative group of nonzero complex numbers $(k = 1, 2, \ldots, l)$. From Theorem 3.4.3. in [66] it follows that this representation of f is unique, and as f is a generalized polynomial hence we have $l = 1$, $m_1 = 1$ and f is a polynomial. The theorem is proved.

Theorem 3.23. *The torsion free rank of any Abelian group is equal to the dimension of the linear space consisting of all complex additive functions of the group in the sense that either both are finite and equal, or both are infinite.*

Proof. Let G be an Abelian group and let let $k = r_0(G) \leq +\infty$. Then G has a subgroup isomorphic to $\mathbb{Z}^k$. If k is infinite, then this is equal to the non-complete direct product of k copies of $\mathbb{Z}$. We will identify this subgroup by $\mathbb{Z}^k$. Obviously $\mathbb{Z}^k$ has at least k linearly independent complex additive functions; for instance we can take the projections onto the different factors of the product group. On the other hand, we have seen in Theorem 3.1 that any homomorphism of a subgroup of an Abelian group into a divisible Abelian group can be extended to a homomorphism of the whole group. As the additive group of complex numbers is obviously divisible, the above mentioned linearly independent complex additive functions of $\mathbb{Z}^k$ can be extended to complex homomorphisms of the whole group G, and the extensions are clearly linearly independent, too. Hence the dimension of the linear space of all complex additive functions of G is not less than the torsion free rank of G.

Now we suppose that $k < +\infty$. Let Φ denote the natural homomorphism of G onto the factor group with respect to $\mathbb{Z}^k$. As it is a torsion group, hence for each element g of G there is a positive integer n such that

$$0 = n\Phi(g) = \Phi(ng),$$

thus ng belongs to the kernel of Φ, which is $\mathbb{Z}^k$. This means that there exist integers $m_1, m_2, \ldots, m_k$ such that

$$ng = (m_1, m_2, \ldots, m_k).$$

Suppose now that there are $k+1$ linearly independent complex additive functions $a_1, a_2, \ldots, a_{k+1}$ on G. Then there exist elements $g_1, g_2, \ldots, g_{k+1}$ in G such that the $(k+1) \times (k+1)$ matrix $(a_i(g_j))$ is regular. For $l = 1, 2, \ldots, k$ we let e_l denote the vector in $\mathbb{C}^k$ whose l-th coordinate is 1, the others are 0. By our above consideration there are integers $m_l^{(j)}$, n_j for $l = 1, 2, \ldots, k$ and $j = 1, 2, \ldots, k+1$ such that

$$n_j g_j = (m_1^{(j)}, m_2^{(j)}, \ldots, m_k^{(j)}).$$

Hence we have

$$a_i(n_j g_j) = a_i(m_1^{(j)}, m_2^{(j)}, \ldots, m_k^{(j)})$$
$$= m_1^{(j)} a_i(e_1) + m_2^{(j)} a_i(e_2) + \cdots + m_k^{(j)} a_i(e_k),$$

and therefore

$$a_i(g_j) = \sum_{l=1}^{k} \frac{m_l^{(j)}}{n_j} a_i(e_l)$$

holds for $i, j = 1, 2, \ldots, k+1$. This means that the linearly independent rows of the matrix $(a_i(g_j))$ are linear combinations of the transpose of the matrix $(a_i(e_l))$ for $l = 1, 2, \ldots, k$ and $i, j = 1, 2, \ldots, k+1$. But this is impossible, because the latter matrix has only k columns, hence its rank is at most k.

We have shown that if the torsion free rank of G is the finite number k, then the dimension of the linear space consisting of all complex additive functions of G is at most k, hence the theorem is proved.

Theorem 3.24. *The torsion free rank of an Abelian group is finite if and only if any complex generalized polynomial on the group is a polynomial.*

Proof. Suppose that the torsion free rank of the Abelian group G is finite. If it is zero, then the statement is obvious, because in that case any complex generalized polynomial on G is a constant. Hence we suppose that the torsion free rank of G is the positive integer k, and then by the previous theorem any complex additive function is a linear combination of some fixed linearly independent complex additive functions $a_1, a_2, \ldots, a_k$. First we describe the general form of the multi-additive functions on G. If $B : G \times G \to \mathbb{C}$ is bi-additive, then $x \mapsto B(x, y)$ is additive for any fixed y in G, hence there are functions $\lambda_1, \lambda_2, \ldots, \lambda_k : G \to \mathbb{C}$ such that

$$B(x, y) = \lambda_1(y)a_1(x) + \lambda_2(y)a_2(x) + \cdots + \lambda_k(y)a_k(x)$$

holds for any x, y in G. The linear independence of the functions $a_1, a_2, \ldots, a_k$ implies that there are elements $g_1, g_2, \ldots, g_k$ in G such that the $k \times k$ matrix $(a_i(g_j))$ is regular. Substituting $x = g_j$ for $j = 1, 2, \ldots, k$ into the above equation we get a linear system of equations from which it is clear that the functions $\lambda_1, \lambda_2, \ldots, \lambda_k$ are linear combinations of the functions $y \mapsto B(g_j, y)$ for $j = 1, 2, \ldots, k$, hence they are additive. This means that these functions are also linear combinations of the functions $a_1, a_2, \ldots, a_k$. Therefore the general form of the bi-additive functions on G is the following:

$$B(x, y) = \sum_{i=1}^{k} \sum_{j=1}^{k} b_{ij} a_i(x) a_j(y),$$

where the b_{ij}'s are complex numbers for $i, j = 1, 2, \ldots, k$. Repeating this argument we get by induction that for any positive integer n the general form of the n-additive functions on G is the following:

$$A(x_1, x_2, \ldots, x_n) = \sum_{i_1=1}^{k} \sum_{i_2=1}^{k} \cdots \sum_{i_n=1}^{k} m_{i_1 i_2 \ldots i_n} a_{i_1}(x_1) a_{i_2}(x_2) \ldots a_{i_n}(x_n),$$

where the $m_{i_1 i_2 \ldots i_n}$'s are complex numbers for $i_1, i_2 \ldots, i_n = 1, 2, \ldots, k$. From this it is clear that on G any complex generalized polynomial is a polynomial.

Conversely, suppose that the torsion free rank of G is infinite. We take a maximal independent system of elements of infinite order. Let $X = \{x_i : i \in i\}$ be such a system. Here I is some infinite set. Then for every i in I there is a homomorphism $p_i : G \to \mathbb{C}$ such that $p_i(x_j) = 1$ for $i = j$ and $p_i(x_j) = 0$ for $i \neq j$. Indeed, first we find such a homomorphism from the subgroup H generated by X then we extend it to G. It follows that the functions p_i are linearly independent for different values i in I. Let finally

$$B(x, y) = \sum_{i \in I} p_i(x) p_i(y)$$

for each x, y in G. The sum is finite for any fixed x, y. Indeed, by the maximality of X, for each g in G there is a positive integer n such that ng belongs to H. Then

$$ng = m_1 x_{j_1} + m_2 x_{j_2} + \cdots + m_k x_{j_k}$$

with some integers $m_1, m_2, \ldots, m_k$ and elements $x_{j_1}, x_{j_2}, \ldots, x_{j_k}$ in X. It is clear that $p_i(g) = 0$ for every $i \neq j_1, j_2, \ldots, j_k$.

Obviously B is a symmetric and bi-additive function. On the other hand, if x_j is any element of X then we have

$$B(x, x_j) = \sum_{i \in I} p_i(x) p_i(x_j) = p_j(x),$$

hence the functions $x \mapsto B(x, x_j)$ are linearly independent for different values i in I. All these functions belong to the translation invariant linear space generated by the function $x \mapsto B(x, x)$, hence this linear space is of infinite dimension. Indeed, constant functions — as second differences of $x \mapsto B(x, x)$ — obviously belong to this space, further

$$B(x + x_j, x + x_j) - B(x, x) - B(x_j, x_j) = 2B(x, x_j)$$

holds for each x in G, and the left hand side belongs to the translation invariant linear space generated by the function $x \mapsto B(x, x)$.

On the other hand, we have seen above that the translation invariant linear space generated by any polynomial is of finite dimension, hence the generalized polynomial $x \mapsto B(x, x)$ is not a polynomial.

Actually we have proved the following theorem.

Theorem 3.25. *The torsion free rank of an Abelian group is finite if and only if any complex bi-additive function is a bilinear function of complex additive functions.*

According to Conjecture 2 about spectral synthesis formulated in Section 3.5, here we give an equivalent formulation of that conjecture: spectral synthesis holds on an Abelian group if and only if any complex bi-additive function on the group is a bilinear function of complex additive functions.

References: [5], [35], [79], [24], [16], [12], [66], [44], [69], [70], [61], [59], [3], [32], [75], [76].

4

Spectral synthesis and functional equations

4.1 Convolution type functional equations

We recall from the previous part that for a given locally compact Abelian group G any proper closed translation invariant subspace of $\mathcal{C}(G)$ is called a variety. The set of all exponentials in a variety is called the *spectrum* of the variety, and the set of all exponential monomials in a variety is called the *spectral set* of the variety. If V is a variety, then $sp\, V$ denotes the spectrum of V and we write $sp\, f$ for $sp\, \tau(f)$. If μ is in $\mathcal{M}_c(G)$, then we use the notation $sp\, \mu$ for the spectrum of the annihilator of the ideal generated by μ, and for any subset Λ of $\mathcal{M}_c(G)$ the spectrum, or spectral set of Λ, is the spectrum, or the spectral set of the annihilator of the ideal generated by Λ.

The convolution between $\mathcal{C}(G)$ and $\mathcal{M}_c(G)$ is defined by the formula

$$f * \mu(x) = \int f(x - y)\, d\mu(y) = \langle \tau_{-x} f^*, \mu \rangle$$

for all f in $\mathcal{C}(G)$, μ in $\mathcal{M}_c(G)$ and x in G. Using the notation μ^* for the measure defined by $\langle f, \mu^* \rangle = \langle f^*, \mu \rangle$, we can also write

$$f * \mu(x) = \langle \tau_x f, \mu^* \rangle$$

for all f in $\mathcal{C}(G)$, μ in $\mathcal{M}_c(G)$ and x in G. Hence $f * \mu = 0$ if and only if f belongs to the annihilator of the ideal generated by μ^*. In other words, the set of all solutions f of the equation

$$f * \mu = 0$$

is identical with the annihilator of the ideal generated by μ^*. More generally, let $\Lambda \neq \{0\}$ be a nonempty set of measures in $\mathcal{M}_c(G)$. Then the system of equations

$$f * \mu = 0$$

for all μ in Λ is called the *system of convolution type equations associated with* Λ. The solution space of this system is obviously a variety. Indeed, it is the intersection of the annihilators of the ideals generated by the elements of $\Lambda^* = \{\mu^* \mid \mu \in \Lambda\}$. It is clear that any variety arises in this manner. Hence the study of varieties in $\mathcal{C}(G)$ is equivalent to the study of systems of convolution type functional equations on G. Spectral analysis for a given variety means that the corresponding system of convolution type functional equations has an exponential solution, that is, the spectrum of the generating set of measures is nonempty. The meaning of spectral synthesis for a given variety is that any solution of the corresponding system of convolution type functional equations can be approximated uniformly on compact sets by linear combinations of exponential monomial solutions of the system, that is, the linear hull of the spectral set of the generating set of measures is dense in the solution space. Consequently, in case of spectral synthesis the solution space of the system can be completely described by the exponential monomial solutions, hence any method for finding exponential monomial solutions of convolution type functional equations is very useful and highly appreciated.

In the applications the following problem arises: if two systems of convolution type functional equations are given, how to decide if one of them implies the other, or how to decide, if they are equivalent? Here "implies" means that any solution of the first one is a solution of the second one, and they are equivalent if they mutually imply each other. If spectral synthesis holds for $\mathcal{C}(G)$, then the problem can be reduced to the study of spectral sets: one system implies the other if and only if the spectral set of the first is included in the spectral set of the second, and two systems are equivalent if and only if their spectral sets are identical. In some cases it is useful to know that the problem of implication and equivalence can be reduced to compactly generated locally compact Abelian groups. Let F be any closed subgroup of G and let Λ be a set of measures in $\mathcal{M}_c(G)$. We define the *restriction of* Λ *to* F as the set of all measures in Λ whose support lies in F. Then the following theorem holds (see also [68]).

Theorem 4.1. *Let G be a locally compact Abelian group and let Λ, Γ be sets of compactly supported complex Radon measures on G. If the restriction of Λ to any compactly generated closed subgroup of G implies the restriction of Γ to the same subgroup, then Λ implies Γ. In particular, Λ and Γ are equivalent if and only if their restrictions to any compactly generated subgroup are equivalent.*

Proof. Suppose that the restriction of Λ to any compactly generated closed subgroup of G implies the restriction of Γ to the same subgroup, and Λ does not imply Γ. Then there exists a function f satisfying $f * \mu = 0$ for all μ in Λ and $f * \gamma_0(x_0) \neq 0$ for some γ_0 in Γ and x_0 in G. Let F denote the closed subgroup generated by x_0 and the support of γ_0. Obviously, the restriction f_F of f to F satisfies $f_F * \mu = 0$ for all μ in Λ with support in F. As F is compactly generated, and the restriction of Λ to F implies the restriction of Γ to the F, we infer that $f_F * \gamma_0(x) = 0$ for all x in F, as γ_0 is in Γ with support in F. But this contradicts the fact that x_0 is in F.

In case of discrete Abelian groups this theorem has the following reformulation.

Theorem 4.2. *Let G be an Abelian group and let Λ, Γ be sets of finitely supported complex measures on G. If the restriction of Λ to any finitely generated subgroup of G implies the restriction of Γ to the same subgroup, then Λ implies Γ. In particular, Λ and Γ are equivalent if and only if their restrictions to any finitely generated subgroup are equivalent.*

Using Theorem 2.23 on spectral synthesis for finitely generated Abelian groups we can apply this theorem immediately for convolution type functional equations on discrete Abelian groups.

Theorem 4.3. *Let G be an Abelian group and let Λ, Γ be sets of finitely supported complex measures on G. Then Λ implies Γ if and only if the spectral set of the restriction of Λ to any finitely generated subgroup is a subset of the spectral set of the restriction of Γ to the same subgroup. In particular, Λ and Γ are equivalent if and only if their spectral sets of their restrictions to any finitely generated subgroup are the same.*

Proof. Suppose that Λ implies Γ and F is any finitely generated subgroup of G. Then the restriction of Λ to F implies the restriction of Γ to F, hence any solution of the restriction of Λ on F is a solution of the restriction of Γ on F. In particular, this applies for exponential monomial solutions, hence the spectral set of the restriction of Λ to F is a subset of the spectral set of the restriction of Γ to F, which proves the necessity. Conversely, we suppose that the spectral set of the restriction Λ_F of Λ to F is a subset of the spectral set of the restriction Γ_F of Γ to F. This means that any exponential monomial in the solution space of Λ_F belongs to the solution space of Γ_F. As these solution spaces are varieties and F is finitely generated, by Theorem 2.23 the solution space of Λ_F is included in the solution space of Γ_F, that is, Λ_F implies Γ_F, hence Theorem 4.2 gives our statement.

This theorem makes it possible to reduce questions on implication and equivalence for convolution type functional equations to the study of their exponential monomial solutions. In the following section we present a typical result of this type.

For the results in this section see [66], [68].

4.2 Mean value type functional equations

In [66] we studied the functional equations of mean-value type

$$\left[\sum_{i=1}^{n}\left(\tau_t^i + \tau_{-t}^i\right)\right]f = 2nf,\qquad(4.1)$$

and

$$\left[\prod_{i=1}^{n}(\tau_t^i + \tau_{-t}^i)\right]f = 2^n f,\qquad(4.2)$$

where an Abelian group G is given, n is a positive integer, $f : G^n \to \mathbb{C}$ is a function and τ_t^i denotes the partial translation operator in the i-th variable with increment t, that is,

$$\tau_t^i f(x_1, x_2, \ldots, x_n) = f(x_1, x_2, \ldots, x_{i-1}, x_i + t, x_{i+1}, \ldots, x_n)$$

holds for $i = 1, 2, \ldots, n$ and for all $x_1, x_2, \ldots, x_n, t$ in G. Equation (4.1) and (4.2) are called *octahedron* and *cube equation*, respectively. For $n = 1$ they coincide and they are equivalent to the Jensen equation. For the history of equations of the above and similar type we refer to [66]. It has been proved (see [37], [60]) that (4.1) implies (4.2) for any n, and (4.2) implies (4.1) for $n \leq 4$. It has been conjectured by D. Z. Djoković and H. Haruki (see [60]) that (4.1) and (4.2) are equivalent for all n. Now we show that the previous results can be applied to prove this conjecture. Namely, in the presence of spectral synthesis we can apply the results of the previous section about the equivalence of systems of convolution type functional equations. In [66] we also considered the characterization problem concerning the solutions of (4.1) and (4.2) in the case $G = \mathbb{R}$. We were able to prove a representation theorem for the locally integrable solutions of (4.1) and (4.2), which states that all locally integrable solutions of (4.1) and (4.2) are linear combinations of the partial derivatives of a special polynomial, which is of degree at most $2n - 1$ in each variable. The method we present here allows us to generalize this theorem by proving that on an arbitrary Abelian group any solution of (4.1) or (4.2) is a polynomial of degree at most $2n - 1$ in each variable. The idea is to reduce the problem to the case $G = \mathbb{Z}^k$ and to study polynomial ideals of differential operators.

In the sequel we shall need the following simple results, the first of which is proved in [66] (Lemma 14.1, p.119.) and the second follows easily by induction.

Theorem 4.4. *Let G be an Abelian group. Then any nonzero complex exponential on G is an extremal point of the convex hull of all nonzero complex exponentials on G.*

Theorem 4.5. *Let* $i_1, i_2, \ldots, i_n$ *be nonnegative integers. Then we have*

$$\sum_{(\varepsilon_1, \varepsilon_2, \ldots, \varepsilon_n) \in \{-1,1\}^n} \varepsilon_1^{i_1} \varepsilon_2^{i_2} \ldots \varepsilon_n^{i_n} = \left(1 + (-1)^{i_1}\right)\left(1 + (-1)^{i_2}\right) \ldots \left(1 + (-1)^{i_n}\right).$$

We note that this statement can be reformulated as follows: the given sum is different from zero if and only if all the exponents $i_1, i_2, \ldots, i_n$ are even, and in this case it is equal to 2^n.

In the sequel we shall use multi-index notation. Multi-indices of the same dimension are added component-wise and ordered lexicographically. Similarly, we order vectors of the same dimensional multi-indices lexicographically, corresponding to the ordering of their components. It is clear that both the ordering of multi-indices and that of the vectors of multi-indices are linear.

Let k be a positive integer and let $\alpha = (\alpha_1, \alpha_2, \ldots, \alpha_k)$ be a k-dimensional multi-index in $\mathbb{N}^k$. As in Section 3.3, for the given k-dimensional vector $x = (x_1, x_2, \ldots, x_k)$ we write $x^\alpha = x_1^{\alpha_1} x_2^{\alpha_2} \ldots x_k^{\alpha_k}$ and the factorial $\alpha!$ of the multi-index α is the product of the factorials of its components. The *height* of the multi-index α is equal to $|\alpha| = \alpha_1 + \alpha_2 + \cdots + \alpha_k$. We call a multi-index *even* if its height is an even number. For any nonzero k-dimensional even multi-index α, let $\Gamma_n(\alpha)$ denote the set of all vectors $(\beta_1, \beta_2, \ldots, \beta_l)$ with $1 \leq l \leq n$, where $\beta_1, \beta_2, \ldots, \beta_l$ are nonzero k-dimensional even multi-indices with $\beta_1 \geq \beta_2 \geq \cdots \geq \beta_l$ and $\beta_1 + \beta_2 + \cdots + \beta_l = \alpha$.

For any nonzero k-dimensional even multi-index α let

$$P_n(\alpha) = \sum \frac{\alpha!}{\beta_1! \beta_2! \ldots \beta_n!} x_1^{\beta_1} x_2^{\beta_2} \ldots x_n^{\beta_n},$$

where $x_1, x_2, \ldots, x_n$ are in $\mathbb{R}^k$, and the summation is extended over all k-dimensional even multi-indices $\beta_1, \beta_2, \ldots, \beta_n$ with $\beta_1 + \beta_2 + \cdots + \beta_n = \alpha$. If any component of some exponent is equal to zero, then the corresponding factor is considered to be 1.

For an arbitrary positive integer l with $1 \leq l \leq n$ and for any nonzero k-dimensional multi-indices $\beta_1, \beta_2, \ldots, \beta_l$ with $\beta_1 \geq \beta_2 \geq \cdots \geq \beta_l$ we denote by $Q_n(\beta_1, \beta_2, \ldots, \beta_l)$ the sum of all different monomials of the form $x_{i_1}^{\beta_1} x_{i_2}^{\beta_2} \ldots x_{i_l}^{\beta_l}$, where $i_1, i_2, \ldots, i_l$ are different integers between 1 and n, and $x_{i_1}, x_{i_2}, \ldots, x_{i_l}$ are in $\mathbb{R}^k$. For instance

$$Q_n(\beta) = \sum_{i=1}^n x_i^\beta.$$

We remark that in the sequel we shall consider the polynomials $P_n(\alpha)$ and $Q_n(\beta_1, \beta_2, \ldots, \beta_l)$ mainly as polynomial differential operators in $n \cdot k$ variables by substituting $x_i = \partial_i = (\partial_{i,1}, \partial_{i,2}, \ldots, \partial_{i,k})$ and by interpreting addition and multiplication in the obvious way.

It is clear that we have the representation

$$P_n(\alpha) = \sum_{l=1}^{n} \sum_{(\beta_1, \beta_2, \ldots, \beta_l) \in \Gamma_n(\alpha)} \lambda_{\beta_1, \beta_2, \ldots, \beta_l} Q_n(\beta_1, \beta_2, \ldots, \beta_l).$$

Here the coefficients are positive integers and the coefficient of $Q_n(\alpha)$ is 1. It is also easy to see that if we put $x_{n+1} = 0$ in $P_{n+1}(\alpha)$, then we get $P_n(\alpha)$, and if we put $x_{n+1} = 0$ in $Q_{n+1}(\beta_1, \beta_2, \ldots, \beta_l)$, then we get 0 for $l = n+1$ and $Q_n(\beta_1, \beta_2, \ldots, \beta_l)$ for $l \leq n$.

We have another representation for $P_n(\alpha)$.

Theorem 4.6. *Let n, k be positive integers and α a k-dimensional multi-index. Then we have*

$$P_n(\alpha) = 2^{-n} \sum_{(\varepsilon_1, \varepsilon_2, \ldots, \varepsilon_n) \in \{-1,1\}^n} \left(\sum_{i=1}^{n} \varepsilon_i x_i \right)^{\alpha}.$$

Proof. We use the Polynomial Theorem and Theorem 4.5 in the following computation:

$$2^{-n} \sum_{(\varepsilon_1, \varepsilon_2, \ldots, \varepsilon_n) \in \{-1,1\}^n} \left(\sum_{i=1}^{n} \varepsilon_i x_i \right)^{\alpha}$$

$$= 2^{-n} \sum_{(\varepsilon_1, \varepsilon_2, \ldots, \varepsilon_n) \in \{-1,1\}^n} \sum_{\beta_1 + \cdots + \beta_n = \alpha} \frac{\alpha!}{\beta_1! \ldots \beta_n!} \varepsilon_1^{|\beta_1|} \ldots \varepsilon_n^{|\beta_n|} x_1^{\beta_1} \ldots x_n^{\beta_n}$$

$$= 2^{-n} \sum_{\beta_1 + \cdots + \beta_n = \alpha} \frac{\alpha!}{\beta_1! \ldots \beta_n!} \left(\sum_{(\varepsilon_1, \varepsilon_2, \ldots, \varepsilon_n) \in \{-1,1\}^n} \varepsilon_1^{|\beta_1|} \ldots \varepsilon_n^{|\beta_n|} \right) x_1^{\beta_1} \ldots x_n^{\beta_n}$$

$$= 2^{-n} \sum_{\beta_1 + \cdots + \beta_n = \alpha} \frac{\alpha!}{\beta_1! \ldots \beta_n!} \left(1 + (-1)^{|\beta_1|}\right) \ldots \left(1 + (-1)^{|\beta_n|}\right) x_1^{\beta_1} \ldots x_n^{\beta_n}$$

$$= \sum_{\beta_1 + \cdots + \beta_n = \alpha, \, \beta_i \text{ is even}} \frac{\alpha!}{\beta_1! \ldots \beta_n!} x_1^{\beta_1} x_2^{\beta_2} \ldots x_n^{\beta_n} = P_n(\alpha).$$

Obviously, any product $Q_n(\beta_1) Q_n(\beta_2) \ldots Q_n(\beta_l)$ with $(\beta_1, \beta_2, \ldots, \beta_l)$ in $\Gamma_n(\alpha)$ is a linear combination of the polynomials $Q_n(\delta_1, \delta_2, \ldots, \delta_j)$ whenever $1 \leq j \leq l$ and $(\delta_1, \delta_2, \ldots, \delta_j)$ is in $\Gamma_n(\alpha)$, and in this combination $Q_n(\alpha)$ has coefficient 1; further any $Q_n(\delta_1, \delta_2, \ldots, \delta_j)$ with nonzero coefficient has the property that $(\delta_1, \delta_2, \ldots, \delta_j) \geq (\beta_1, \beta_2, \ldots, \beta_l)$. We remark that if $j < l$, then the vector $(\delta_1, \delta_2, \ldots, \delta_j)$ is "shorter" than $(\beta_1, \beta_2, \ldots, \beta_l)$, hence in order to compare them we add zero components to it. We do the same always when ordering vectors with different numbers of components.

Now we fix a nonzero k-dimensional even multi-index α. Let $\mathcal{A}$ denote the linear hull of the polynomials $Q_n(\beta_1, \beta_2, \ldots, \beta_l)$ with $(\beta_1, \beta_2, \ldots, \beta_l)$ in $\Gamma_n(\alpha)$. Obviously this set of polynomials is linearly independent, as the functions $Q_n(\beta_1, \beta_2, \ldots, \beta_l)$ are different monomials in $n \cdot k$ variables of the same degree for different choices of $(\beta_1, \beta_2, \ldots, \beta_l)$. Let further $\mathcal{B}$ denote the linear hull of the polynomials $P_n(\alpha)$ and $Q_n(\beta_1)Q_n(\beta_2) \ldots Q_n(\beta_l)$ with $l \geq 2$ and $(\beta_1, \beta_2, \ldots, \beta_l)$ in $\Gamma_n(\alpha)$.

Theorem 4.7. *With the above notation $\mathcal{A} = \mathcal{B}$.*

Proof. By the above remarks it is clear that $\mathcal{B} \subseteq \mathcal{A}$. For the converse we observe first that the cardinalities of the two given sets of polynomials generating $\mathcal{A}$ and $\mathcal{B}$ are the same, therefore it is enough to prove that the given polynomials generating $\mathcal{B}$ are linearly independent. The polynomials $Q_n(\beta_1)Q_n(\beta_2) \ldots Q_n(\beta_l)$ with $l \geq 2$ and $(\beta_1, \beta_2, \ldots, \beta_l)$ in $\Gamma_n(\alpha)$ are obviously linearly independent, hence it is enough to show that $P_n(\alpha)$ is not a linear combination of them. If we suppose the contrary for some n, then according to a previous remark, by substituting $x_n = 0$ we get that it is also the case for $n - 1$. Hence it is enough to show that the given polynomials generating $\mathcal{B}$ are linearly independent for $n = 2$. We have seen above that the polynomials $P_2(\alpha)$ and $Q_2(\beta_1)Q_2(\beta_2) \ldots Q_2(\beta_l)$ with $l \geq 2$ and $(\beta_1, \beta_2, \ldots, \beta_l)$ in $\Gamma_2(\alpha)$ can uniquely be written as linear combinations of the polynomials $Q_2(\beta_1, \beta_2, \ldots, \beta_l)$ with $1 \leq l \leq 2$ and $(\beta_1, \beta_2, \ldots, \beta_l)$ in $\Gamma_2(\alpha)$. We show that the quadratic matrix of this linear transformation is regular. Suppose that the first row contains the coefficients of the linear expression for $P_2(\alpha)$ in terms of the polynomials $Q_2(\beta_1, \beta_2, \ldots, \beta_l)$ corresponding to the decreasing order of $(\beta_1, \beta_2, \ldots, \beta_l)$, hence the first row has the form $(1, \lambda_1, \lambda_2, \ldots, \lambda_N)$, where $N \geq 2$ is an integer, and $\lambda_1, \lambda_2, \ldots, \lambda_N$ are positive integers. The second, third, etc., rows contain the coefficients of the linear expressions for the polynomials $Q_2(\beta_1)Q_2(\beta_2) \ldots Q_2(\beta_l)$ with $l = 2$ corresponding to the decreasing order of $(\beta_1, \beta_2, \ldots, \beta_l)$. We have two cases. If the multi-index α is not of the form $\beta + \beta$ with some (β, β) in $\Gamma_2(\alpha)$, then $Q_2(\beta_1)Q_2(\beta_2) = Q_2(\alpha) + Q_2(\beta_1, \beta_2)$ for any (β_1, β_2) in $\Gamma_2(\alpha)$, hence any row, which is different from the first one, has only two nonzero entries, which are equal to 1, namely, the second row is $(1, 1, 0, \ldots, 0)$, the third is $(1, 0, 1, 0, \ldots, 0)$, and the last one is $(1, 0, \ldots, 0, 1)$, where the dots represent zeros. The determinant of this matrix is $1 - \lambda_1 - \lambda_2 - \cdots - \lambda_N$, which is different from zero. In the second case $\alpha = \beta + \beta$ for some β, and in this case $Q_2(\beta)Q_2(\beta) = Q_2(\alpha) + 2Q_2(\beta, \beta)$, which means that in the corresponding row of the matrix the first entry is 1 and the other nonzero entry is 2, instead of 1. Multiplying the corresponding column of the matrix by $\frac{1}{2}$, the first row changes to $(1, \lambda_1, \ldots, \frac{\lambda_i}{2}, \ldots, \lambda_N)$ and the determinant to $1 - \lambda_1 - \cdots - \frac{\lambda_i}{2} - \cdots - \lambda_N$, which is also different from zero. This means that the matrix of the linear transformation, which maps a basis of $\mathcal{A}$ to a generating set of $\mathcal{B}$, is regular, hence the given generating set of $\mathcal{B}$ is linearly independent and $\mathcal{A} \subseteq \mathcal{B}$. The theorem is proved.

For any nonzero k-dimensional multi-index α let $I_n(\alpha)$, respectively $J_n(\alpha)$ denote the ideal generated by all the polynomials $Q_n(\beta)$, respectively $P_n(\beta)$ in the ring of complex polynomials in $n \cdot k$ variables, where $\beta \leq \alpha$ is nonzero and even. Obviously $I_n(\beta) \subseteq I_n(\alpha)$ and $J_n(\beta) \subseteq J_n(\alpha)$ for $\beta \leq \alpha$.

Theorem 4.8. *The ideals $I_n(\alpha)$ and $J_n(\alpha)$ are identical.*

Proof. If the height of α is 2, then $I_n(\alpha)$ is generated by $Q_n(\alpha)$, and $J_n(\alpha)$ is generated by $P_n(\alpha)$, which are equal, hence $I_n(\alpha) = J_n(\alpha)$. Suppose that we have proved the theorem for all nonzero k-dimensional multi-indices with height less than $2N$ and let $|\alpha| = 2N$. We have seen above that $\mathcal{A} \subseteq \mathcal{B}$, hence $Q_n(\alpha)$ is a linear combination of the polynomials $P_n(\alpha)$ and $Q_n(\beta_1)Q_n(\beta_2)\dots Q_n(\beta_l)$ with $2 \leq l \leq n$ and $(\beta_1, \beta_2, \dots, \beta_l)$ in $\Gamma_n(\alpha)$. As $\beta_1 + \beta_2 + \dots + \beta_l = \alpha$, here the height of β_i is less than $2N$, hence by our assumption $Q_n(\beta_i)$ is in $J_n(\beta_i) \subseteq J_n(\alpha)$ for $i = 1, 2, \dots, l$. Since $P_n(\alpha)$ also belongs to $J_n(\alpha)$ we infer that $Q_n(\alpha)$ belongs to $J_n(\alpha)$, and $I_n(\alpha) \subseteq J_n(\alpha)$. Conversely, the polynomials $Q_n(\beta_1, \beta_2, \dots, \beta_l)$ with $2 \leq l \leq n$ and $(\beta_1, \beta_2, \dots, \beta_l)$ in $\Gamma_n(\alpha)$ are linear combinations of products of the form $Q_n(\delta_1)Q_n(\delta_2)\dots Q_n(\delta_j)$ with $2 \leq j \leq n$ and $(\delta_1, \delta_2, \dots, \delta_j)$ in $\Gamma_n(\alpha)$, which all belong to $I_n(\alpha)$. Further, $P_n(\alpha)$ is a linear combination of $Q_n(\alpha)$ and $Q_n(\beta_1, \beta_2, \dots, \beta_l)$ with $2 \leq l \leq n$ and $(\beta_1, \beta_2, \dots, \beta_l)$ in $\Gamma_n(\alpha)$, which all belong to $I_n(\alpha)$. This implies $J_n(\alpha) \subseteq I_n(\alpha)$, and our statement is proved.

Theorem 4.9. *The polynomial solutions of (4.1) and (4.2) are identical if $G = \mathbb{Z}^k$.*

Proof. Any polynomial solution of (4.1) or (4.2) on $\mathbb{Z}^k$ is a complex polynomial in $n \cdot k$ variables. If (4.1) or (4.2) holds for a polynomial, then fixing $x_1, x_2, \dots, x_n$ in $\mathbb{Z}^k$ we have a polynomial identity in the variable t in $\mathbb{Z}^k$, which must hold for all $x_1, x_2, \dots, x_n$ and t in $\mathbb{R}^k$, too. For any fixed $x_1, x_2, \dots, x_n$ the two polynomials in $t = (t_1, t_2, \dots, t_k)$ on the two sides of (4.1) and (4.2) have the same value at $t = (0, 0, \dots, 0)$, hence they are identical if and only if their derivatives of all order are equal at $t = (0, 0, \dots, 0)$, by the Taylor Formula. Let α be any nonzero k-dimensional multi-index. Applying the differential operator $\partial_t^\alpha = \partial_{t_1}^{\alpha_1} \partial_{t_2}^{\alpha_2} \dots \partial_{t_k}^{\alpha_k}$ on both sides of (4.1) and then substituting $t = (0, 0, \dots, 0)$, we have that a necessary and sufficient condition for the polynomial $f : \left(\mathbb{Z}^k\right)^n \to \mathbb{C}$ is a solution of (4.1) is that

$$\left(1 + (-1)^{|\alpha|}\right) \sum_{i=1}^{n} \partial_i^\alpha f = 0 \,.$$

Here $\partial_i^\alpha = \partial_{i,1}^{\alpha_1} \partial_{i,2}^{\alpha_2} \dots \partial_{i,k}^{\alpha_k}$, where $\partial_{i,j}$ denotes partial differentiation with respect to the j-th component of the i-th variable for $i = 1, 2, \dots, n$ and

$j = 1, 2, \ldots, k$. This means that the polynomial $f : \left(\mathbb{Z}^k\right)^n \to \mathbb{C}$ satisfies (4.1) if and only if for any nonzero k-dimensional even multi-index α,

$$\sum_{i=1}^{n} \partial_i^{\alpha} f = 0 \,,$$

or

$$Q_n(\alpha)f = 0 \tag{4.3}$$

holds, where $Q_n(\alpha)$ is the polynomial differential operator, obtained as above with $x_i = \partial_i$.

Now we apply the differential operator ∂_t^{α} on both sides of (4.2) and substitute $t = (0, 0, \ldots, 0)$. Then we have that the polynomial $f : \left(\mathbb{Z}^k\right)^n \to \mathbb{C}$ satisfies (4.2) if and only if for any nonzero k-dimensional even multi-index α

$$\sum_{(\varepsilon_1, \varepsilon_2, \ldots, \varepsilon_n) \in \{-1, 1\}^n} \left(\sum_{i=1}^{n} \varepsilon_i \partial_i\right)^{\alpha} f = 0 \,,$$

or

$$2^n P_n(\alpha)f = 0 \tag{4.4}$$

holds, where $P_n(\alpha)$ is the polynomial differential operator, obtained as above with $x_i = \partial_i$. By Theorem 4.8 the ideals generated by the polynomials $Q_n(\alpha)$ and $P_n(\alpha)$ are identical, that is, the systems of partial differential equations (4.3) and (4.4) are equivalent. Hence the polynomial solutions of (4.1) and (4.2) are identical if $G = \mathbb{Z}^k$.

Theorem 4.10. *The functional equations* (4.1) *and* (4.2) *are equivalent on any Abelian group G for each positive integer n.*

Proof. By Theorem 4.2 it is enough to show that the restrictions of (4.1) and (4.2) to any finitely generated subgroup of G are equivalent, that is, (4.1) and (4.2) is equivalent on any finitely generated Abelian group. Using the result of Lefranc's theorem 2.22 on spectral synthesis for $\mathbb{Z}^k$, we show that (4.1) and (4.2) are equivalent if $G = \mathbb{Z}^k$. By Theorem 4.3 it is enough to show, that in this case the exponential monomial solutions of (4.1) and (4.2) are identical. However, if an exponential monomial has the form pm, where $p : G^m \to \mathbb{C}$ is a polynomial, and $m : G^n \to \mathbb{C}$ is an exponential and it is a solution of (4.1) or (4.2), then the exponential m is also a solution of (4.1) or (4.2), because the solution spaces of these equations are translation invariant linear function spaces closed under pointwise convergence. In this case Lemma 4.2 in [66], p. 40. can be applied. If $m : G^n \to \mathbb{C}$ is an exponential, then it has the form $m(x_1, x_2, \ldots, x_n) = m_1(x_1)m_2(x_2) \ldots m_n(x_n)$, where $m_1, m_2, \ldots, m_n : G \to \mathbb{C}$ are exponentials. Substituting m into (4.1) or (4.2) we get immediately by Theorem 4.4 that $m_1 = m_2 = \cdots = m_n = 1$, hence

$m = 1$, that is, any exponential monomial solution of (4.1) or (4.2) is a polynomial. By Theorem 4.9 the polynomial solutions of (4.1) and (4.2) are identical if $G = \mathbb{Z}^k$.

Suppose now that G is an arbitrary finitely generated Abelian group, and let $\varphi : \mathbb{Z}^k \to G$ be a surjective homomorphism, where k is some positive integer. The function $\Phi : (\mathbb{Z}^k)^n \to G^n$ defined by

$$\Phi(z_1, z_2, \ldots, z_n) = (\varphi(z_1), \varphi(z_2), \ldots, \varphi(z_n))$$

for $z_1, z_2, \ldots, z_n$ in $\mathbb{Z}^k$ is a surjective homomorphism. If $f : G^n \to \mathbb{C}$ is a solution of (4.1) on G^n, then $f \circ \Phi : (\mathbb{Z}^k)^n \to \mathbb{C}$ is obviously a solution of (4.1) on $(\mathbb{Z}^k)^n$, hence $f \circ \Phi$ satisfies (4.2) on $(\mathbb{Z}^k)^n$, which implies that $f : G^n \to \mathbb{C}$ is a solution of (4.2) on G^n. The converse follows in the same manner, hence our theorem is proved.

Theorem 4.11. *If $n, k \geq 1$ are arbitrary integers, then each polynomial solution $f : (\mathbb{Z}^k)^n \to \mathbb{C}$ of (4.1) satisfies*

$$\partial_i^\alpha f = 0$$

for any k-dimensional multi-index α with $|\alpha| = 2n$ $(i = 1, 2, \ldots, n)$.

Proof. We prove the statement for $i = 1$. First we show by induction on l that

$$\partial_1^\alpha \left(\sum_{1 < i_1 < \cdots < i_{n-l} \leq n} \partial_{i_1}^{e_p + e_q} \ldots \partial_{i_{n-l}}^{e_p + e_q} \right) f = 0 \qquad (4.5)$$

holds for any k-dimensional multi-index α with $|\alpha| = 2l$ $(l = 1, 2, \ldots)$. Here e_p denotes the k-dimensional multi-index, whose p-th component is 1 and all the other components are 0. If $l = 1$ and $\beta_1 = \beta_2 = \cdots = \beta_n = e_p + e_q$, then

$$\partial_1^{e_p + e_q} \left(\sum_{1 < i_1 < \cdots < i_{n-1} \leq n} \partial_{i_1}^{e_p + e_q} \ldots \partial_{i_{n-1}}^{e_p + e_q} \right) f = Q_n(\beta_1, \beta_2, \ldots, \beta_n) f = 0 \,,$$

by Theorem 4.7, which is (4.5) for $l = 1$. Suppose that (4.5) holds for l and let β be a k-dimensional multi-index with $|\beta| = 2(l+1)$. Then β has the form $\beta = \alpha + e_p + e_q$ with some $1 \leq p, q \leq k$, where $|\alpha| = 2l$. By Theorem 4.7 we have

$$Q_n(\beta_1, \beta_2, \ldots, \beta_{n-l}) f = 0 \qquad (4.6)$$

with $\beta_1 = \beta_2 = \cdots = \beta_{n-l} = e_p + e_q$, that is,

$$\partial_1^{e_p + e_q} \left(\sum_{1 < i_1 < \cdots < i_{n-l-1} \leq n} \partial_{i_1}^{e_p + e_q} \ldots \partial_{i_{n-l-1}}^{e_p + e_q} \right) f$$

$$+ \left(\sum_{1 < i_1 < \cdots < i_{n-l} \leq n} \partial_{i_1}^{e_p + e_q} \ldots \partial_{i_{n-l}}^{e_p + e_q} \right) f = 0.$$

(Here in the first term we collected all differential operators which contain $\partial_1^{e_p+e_q}$ as a factor, and the second term is the remaining part.) Applying ∂_1^α on both sides we have

$$
0 = \partial_1^\beta \left(\sum_{1 < i_1 < \cdots < i_{n-l-1} \leq n} \partial_{i_1}^{e_p+e_q} \ldots \partial_{i_{n-l-1}}^{e_p+e_q} \right) f
$$
$$
+ \partial_1^\alpha \left(\sum_{1 < i_1 < \cdots < i_{n-l} \leq n} \partial_{i_1}^{e_p+e_q} \ldots \partial_{i_{n-l}}^{e_p+e_q} \right) f
$$
$$
= \partial_1^\beta \left(\sum_{1 < i_1 < \cdots < i_{n-l-1} \leq n} \partial_{i_1}^{e_p+e_q} \ldots \partial_{i_{n-l-1}}^{e_p+e_q} \right) f,
$$

hence (4.5) holds with β in place of α. Finally, by the substitution $l = n$ in (4.5) we have our statement.

Theorem 4.12. *Let G be an Abelian group and n a positive integer. Then any complex valued solution of (4.1) or (4.2) is a polynomial of degree at most $2n - 1$ in each variable.*

Proof. By Theorem 4.10, equations (4.1) and (4.2) are equivalent on any Abelian group hence it is enough to deal with (4.1). Let k be any positive integer and let $f : \left(\mathbb{Z}^k \right)^n \to \mathbb{C}$ be a polynomial solution of (4.1); then by Theorem 4.11,

$$
\partial_i^\alpha f = 0
$$

holds for any k-dimensional multi-index α with $|\alpha| = 2n$ and for $i = 1, 2, \ldots, n$. If $\Delta_{i,t}$ denotes the partial difference operator with increment t in the i-th variable on functions $f : \left(\mathbb{Z}^k \right)^n \to \mathbb{C}$ for $i = 1, 2, \ldots, n$, then this implies that

$$
\Delta_{i,t}^{2n} f(z_1, z_2, \ldots, z_n) = 0 \qquad (4.7)
$$

holds for any $t, z_1, z_2, \ldots, z_n$ in $\mathbb{Z}^k$, that is, the function

$$
z \mapsto f(z_1, z_2, \ldots, z_{i-1}, z, z_{i+1}, \ldots, z_n)
$$

is a polynomial of degree at most $2n-1$ for fixed $z_1, z_2, \ldots, z_{i-1}, z, z_{i+1}, \ldots, z_n$ in $\mathbb{Z}^k$, and for $i = 1, 2, \ldots, n$. In other words, f is a polynomial of degree at most $2n - 1$ in each variable. In the proof of Theorem 4.10 we have seen that any exponential monomial solution of (4.1) is actually a polynomial, and now we know that it is of degree at most $2n - 1$ in each variable. It is clear that pointwise limits of nets of such functions are also polynomials of degree at most $2n - 1$ in each variable, hence by Lefranc's theorem 2.22 on spectral synthesis for $\mathbb{Z}^k$ any complex-valued solution of (4.1) on $\left(\mathbb{Z}^k \right)^n$ is a polynomial of degree at most $2n - 1$ in each variable.

The case of finitely generated groups can be treated exactly in the same way, as in Theorem 4.10. Let G be any finitely generated Abelian group and let $\varphi : \mathbb{Z}^k \rightarrow G$ be a surjective homomorphism, where k is some positive integer. The function $\Phi : (\mathbb{Z}^k)^n \rightarrow G^n$ defined by

$$\Phi(z_1, z_2, \ldots, z_n) = (\varphi(z_1), \varphi(z_2), \ldots, \varphi(z_n))$$

for $z_1, z_2, \ldots, z_n$ in $\mathbb{Z}^k$ is a surjective homomorphism. If $f : G^n \rightarrow \mathbb{C}$ is a solution of (4.1) on G^n, then $f \circ \Phi : (\mathbb{Z}^k)^n \rightarrow \mathbb{C}$ is obviously a solution of (4.1) on $(\mathbb{Z}^k)^n$, hence $f \circ \Phi$ satisfies

$$\Delta_{i,t}^{2n}(f \circ \Phi)(z_1, z_2, \ldots, z_n) = 0$$

for all $i = 1, 2, \ldots, n$ and $t, z_1, z_2, \ldots, z_n$ in $\mathbb{Z}^k$. This implies that f satisfies

$$\Delta_{i,s}^{2n} f(x_1, x_2, \ldots, x_n) = 0$$

for all $i = 1, 2, \ldots, n$ and $s, x_1, x_2, \ldots, x_n$ in G, that is, f is a polynomial of degree at most $2n - 1$ in each variable.

Finally, suppose that G is an arbitrary Abelian group. Let $\{G_\gamma\}$ be the (inductive) set of all finitely generated subgroups of G. For any solution $f : G^n \rightarrow \mathbb{C}$ of (4.1) the function f_γ, which is equal to f on G_γ and zero outside, is a solution of (4.1) on the finitely generated subgroup G_γ, hence f_γ is a polynomial of degree at most $2n - 1$ in each variable on G_γ. As f is the pointwise limit of the net $\{f_\gamma\}$ we infer that f is a polynomial of degree at most $2n - 1$ in each variable, and our theorem is proved.

For the results in this section see [66], [71].

4.3 A functional equation in digital filtering

Spatial filtering is usually performed for deblurring, smoothing, sharpening and enhancing of images. Thus spatial filtering is a technique for modifying or enhancing an image. A particular filter rarely produces a good result for all images. For some images a filter will produce a better result in terms of retaining certain basic information. For other images it may not yield a better result. Based on experience the following mathematical model can be used. Let $\mathbb{Z}_n$ denote the Abelian group of integers modulo n for $n \geq 2$. Then a *digital image* is represented by a function $f : \mathbb{Z}_n \times \mathbb{Z}_n \rightarrow \mathbb{Z}_m$, where n, m are positive integers. Here m is a parameter that indicates the number of gray levels used in representing an image. If the function f denotes an image, then a spatial filtering of f would be

$$\widetilde{f}(x, y) = f(x + t, y + t) + f(x - t, y) + f(x, y - t) \qquad (4.8)$$
$$- f(x - t, y - t) - f(x, y + t) - f(x + t, y)$$

for all x, y, t in $\mathbb{Z}_n$. Here the value of $\tilde{f}(x, y)$ represents the lightness of the image at the point (x, y). The filtered image $\tilde{f}$ retains certain information contained in the original image.

It seems to be reasonable to know on what kind of images a particular filter gives the worst kind of result (that is, it makes the image completely dark). To determine the class of such images with respect to the filter described in (4.8) we encounter the following functional equation:

$$f(x + t, y + t) + f(x - t, y) + f(x, y - t) \qquad (4.9)$$
$$= f(x - t, y - t) + f(x, y + t) + f(x + t, y)$$

for all x, y, t in $\mathbb{Z}_n$, where $n \geq 2$. This functional equation is obviously of convolution type. In [52] the general complex-valued solution of (4.9) has been studied, when (4.9) holds for all x, y, t in a 2-divisible Abelian group G and using spectral synthesis technique it has been shown that all solutions of (4.9) are of the form

$$f(x, y) = B(x, y) + \varphi(x) + \psi(y) + \chi(x - y),$$

where $B : G \times G \to \mathbb{C}$ is bi-additive and $\varphi, \psi, \chi : G \to \mathbb{C}$ are arbitrary functions. The 2-divisibility condition was needed to obtain the above solution. Since now in our case the domain of a digital image is $\mathbb{Z}_n \times \mathbb{Z}_n$, it is appropriate to determine the solution of the functional equation (4.8), when (4.8) holds for all x, y, t in $\mathbb{Z}_n$. As $\mathbb{Z}_n$ is not a 2-divisible group, the solution of (4.8) can not be extracted from that case. First we will illustrate how the above presented methods can be applied in a simple case, where classical Fourier transformation is available. Then we consider another interesting case, the case of integers, where 2-divisibility is still not available, however, spectral synthesis helps again.

If G is any Abelian group, then the function $x \mapsto \chi(-x)$ will be denoted by $\check{\chi}(x)$. It is clear that for any character χ the function $\check{\chi}$ is a character; further a product of characters is a character, too. If G is a finite group, then $|G|$ denotes the number of elements of G. First we observe that if G is a finite Abelian group and $\chi : G \to \mathbb{C}$ is a character, then the value of $\sum_{x \in G} \chi(x)$ is equal to $|G|$ or 0, depending on if χ is identically 1 or not identically 1. Indeed, if χ is identically 1, then the statement is obvious. On the other hand, if y is in G with $\chi(y) \neq 1$, then we have

$$\sum_{x \in G} \chi(x) = \sum_{x \in G} \chi(x + y) = \chi(y) \cdot \sum_{x \in G} \chi(x),$$

which implies our statement.

One can also utilize the simple fact that if G is a finite Abelian group, then different characters of G are linearly independent. For the proof we suppose that

$$\lambda_1 \chi_1(x) + \lambda_2 \chi_2(x) + \cdots + \lambda_k \chi_k(x) = 0 \qquad (4.11)$$

holds for any x in G with different characters $\chi_1, \chi_2, \ldots, \chi_k : G \to \mathbb{C}$ and with some complex numbers $\lambda_1, \lambda_2, \ldots, \lambda_k$. Multiplying (4.11) by $\check{\chi}_j(x)$ for $j = 1, 2, \ldots, k$ and summing up for x in G, by the above remarks we have

$$0 = \sum_{x \in G} \sum_{i=1}^{k} \lambda_i [\chi_i \check{\chi}_j](x) = \sum_{i=1}^{k} \lambda_i \sum_{x \in G} [\chi_i \check{\chi}_j](x) = \lambda_j |G|,$$

and this is our statement.

Our third observation is a simple consequence of these facts. Namely, if G is a finite Abelian group and $\chi_1, \chi_2, \ldots, \chi_k : G \to \mathbb{C}$ are different characters with $\chi_i \neq \check{\chi}_j$ for $i, j = 1, 2 \ldots, k$, then their odd parts are linearly independent. We call the function $x \mapsto \frac{1}{2}(f(x) - f(-x))$ the *odd part* of f. To prove this statement, first we observe that if $\chi_i \neq \chi_j$ and $\chi_i \neq \check{\chi}_j$, then by the previous considerations it follows that

$$\sum_{x \in G} [(\chi_i - \check{\chi}_i)(\check{\chi}_j - \chi_j)](x)$$

$$= \sum_{x \in G} [\chi_i \check{\chi}_j](x) - \sum_{x \in G} [\chi_i \chi_j](x) - \sum_{x \in G} [\check{\chi}_i \check{\chi}_j](x) + \sum_{x \in G} [\check{\chi}_i \chi_j](x) = 0.$$

Now suppose that

$$\lambda_1 [\chi_1 - \check{\chi}_1](x) + \lambda_2 [\chi_2 - \check{\chi}_2](x) + \cdots + \lambda_k [\chi_k - \check{\chi}_k](x) = 0 \qquad (4.12)$$

holds for all x in G with some complex numbers $\lambda_1, \lambda_2, \ldots, \lambda_k$. Similarly, as above, multiplying (4.12) by $[\chi_j - \check{\chi}_j](x)$ for $j = 1, 2 \ldots, k$ and summing up for x in G we have

$$0 = \sum_{x \in G} \sum_{i=1}^{k} \lambda_i [(\chi_i - \check{\chi}_i)(\check{\chi}_j - \chi_j)](x) = \sum_{i=1}^{k} \lambda_i \sum_{x \in G} [(\chi_i - \check{\chi}_i)(\check{\chi}_j - \chi_j)](x)$$

$$= \lambda_j \sum_{x \in G} [(\chi_j - \check{\chi}_j)(\check{\chi}_j - \chi_j)](x) = 2\lambda_j |G|,$$

which gives our statement.

Now we can characterize filters giving completely black images.

Theorem 4.13. *Let $n \geq 2$ be an integer and let $f : \mathbb{Z}_n \times \mathbb{Z}_n \to \mathbb{C}$ be a function satisfying the functional equation (4.9) for all x, y, t in $\mathbb{Z}_n$. Then*

$$f(x, y) = \varphi(x) + \psi(y) + \chi(x - y) \qquad (4.13)$$

holds for all x, y in $\mathbb{Z}_n$, where $\varphi, \psi, \chi : \mathbb{Z}_n \to \mathbb{C}$ are arbitrary functions. Conversely, any function of the form (4.13) is a solution of (4.9).

Proof. It is well known that the dual group of $\mathbb{Z}_n \times \mathbb{Z}_n$ can be identified with $\widehat{\mathbb{Z}}_n \times \widehat{\mathbb{Z}}_n$. Further, the characters of $\mathbb{Z}_n$ are the functions $x \mapsto \xi^{kx}$ ($k = 0, 1, \ldots, n-1$), where ξ is any primitive n-th root of unity. The Fourier transform $\hat{f}$ of f on $\widehat{\mathbb{Z}}_n \times \widehat{\mathbb{Z}}_n$ is defined as

$$\hat{f}(\xi^k, \eta^l) = \sum_{x=0}^{n-1}\sum_{y=0}^{n-1} f(x,y)\xi^{-kx}\eta^{-ly}.$$

The Inversion Formula states that

$$f(x,y) = \frac{1}{n^2}\sum_{k=0}^{n-1}\sum_{l=0}^{n-1}\hat{f}(\xi^k, \eta^l)\xi^{kx}\eta^{ly}.$$

To solve (4.9), we apply Fourier transformation on both sides of (4.9) as functions of (x, y), where t is fixed in $\mathbb{Z}_n$. Then it follows that

$$\hat{f}(\xi^k, \eta^l)\left[\xi^{kt}\eta^{lt} + \xi^{-kt} + \eta^{-lt} - \xi^{-kt}\eta^{-lt} - \xi^{kt} - \eta^{lt}\right] = 0$$

holds for $k, l, t = 0, 1, \ldots, n-1$. Suppose that $\hat{f}(\xi^k, \eta^l) \neq 0$ for some fixed k, l. Then the expression in the brackets vanishes for all t in $\mathbb{Z}_n$. This means that the odd parts of the characters $t \mapsto \xi^{kt}\eta^{lt}$, $t \mapsto \xi^{kt}$ and $t \mapsto \eta^{lt}$ are linearly dependent, hence by our previous remarks we have the following possibilities:

1. $\xi^k = 1$, η is arbitrary,
2. $\eta^l = 1$, ξ is arbitrary,
3. $\xi^k = \eta^l = -1$,
4. $\xi^k = \eta^{-l}$.

According to the Inversion Formula we have

$$f(x,y) = \varphi(x) + \psi(y) + \lambda_0(-1)^{x+y} + \sum_{i=1}^{N}\lambda_i\xi_i^{x-y} \tag{4.14}$$

for all x, y in $\mathbb{Z}_n$, where $\varphi, \psi : \mathbb{Z}_n \to \mathbb{C}$ are arbitrary functions, N is a positive integer, $\lambda_0, \lambda_1, \ldots, \lambda_N$ are complex numbers and $\xi_1, \xi_2, \ldots, \xi_N$ are n-th complex unit roots. As $(-1)^{x+y} = (-1)^{x-y}$, hence f has the form (4.13). The converse statement can be verified by direct computation.

Now we consider the functional equation (4.9) on $\mathbb{Z} \times \mathbb{Z}$. The following two simple results will be needed to establish the result.

Theorem 4.14. *Let $p : \mathbb{R} \times \mathbb{R} \to \mathbb{C}$ be a polynomial satisfying the functional equation*

$$p(x+t, y+t) = p(x-t, y-t) \tag{4.15}$$

for all x, y, t in $\mathbb{Z}$. Then there exists a polynomial $c : \mathbb{R} \to \mathbb{C}$ such that

$$p(x, y) = c(x - y) \tag{4.16}$$

holds for all x, y in $\mathbb{R}$.

Proof. It is clear that if equation (4.15) holds for all x, y, t in $\mathbb{Z}$, then it holds for all x, y, t in $\mathbb{R}$. For fixed x, y in $\mathbb{R}$ differentiating both sides of (4.15) with respect to t and then substituting $t = 0$ we get the first-order partial differential equation

$$\partial_1 p(x, y) + \partial_2 p(x, y) = 0 \qquad (4.17)$$

for all x, y in $\mathbb{R}$. It is easy to see that the general solution of (4.17) has the form (4.16) with some differentiable function $c : \mathbb{R} \to \mathbb{C}$, which — in our case — must be a polynomial.

Theorem 4.15. *Let* $p : \mathbb{R} \times \mathbb{R} \to \mathbb{C}$ *be a polynomial satisfying the functional equation*

$$\begin{aligned} p(x+t, y+t) + p(x-t, y) &+ p(x, y-t) \qquad (4.18) \\ &= p(x-t, y-t) + p(x, y+t) + p(x+t, y) \end{aligned}$$

for all x, y, t *in* $\mathbb{Z}$. *Then there exist polynomials* $a, b, c : \mathbb{R} \to \mathbb{C}$ *such that*

$$p(x, y) = a(x) + b(y) + c(x - y)$$

holds for all x, y *in* $\mathbb{R}$.

Proof. As above, if (4.18) holds for all x, y, t in $\mathbb{Z}$, then it also holds for all x, y, t in $\mathbb{R}$. Then for any fixed x, y in $\mathbb{R}$ we differentiate both sides of (4.18) three times with respect to t, then substituting $t = 0$ we have

$$\partial_1^2 \partial_2 p(x, y) + \partial_1 \partial_2^2 p(x, y) = 0$$

for all x, y in $\mathbb{R}$. Using the notation $q(x, y) = \partial_1 p(x, y) + \partial_2 p(x, y)$ for all x, y in $\mathbb{R}$ it follows that the polynomial $q : \mathbb{R} \times \mathbb{R} \to \mathbb{C}$ satisfies the partial differential equation

$$\partial_1 \partial_2 q(x, y) = 0 \qquad (4.19)$$

for all x, y in $\mathbb{R}$. It is well known, that the general solution of (4.19) has the form $q(x, y) = A(x) + B(y)$ for all x, y in $\mathbb{R}$, where $A, B : \mathbb{R} \to \mathbb{C}$ are differentiable functions. As q is a polynomial, A, B must be polynomials, too. Now we choose polynomials $a, b : \mathbb{R} \to \mathbb{C}$ such that $A = a'$, $B = b'$, where a' and b' denote the derivatives of a and b, respectively. If

$$P(x, y) = p(x, y) - a(x) - b(y) \qquad (4.20)$$

for all x, y in $\mathbb{R}$, then by (4.19) the polynomial P satisfies the partial differential equation

$$\partial_1 P(x, y) + \partial_2 P(x, y) = 0$$

for all x, y in $\mathbb{R}$. As above, the solution of the last equation is of the form $P(x, y) = c(x - y)$ with some differentiable function $c : \mathbb{R} \to \mathbb{C}$, which — in our case — must be a polynomial. From (4.20) our statement follows.

Now we establish our result which depends on spectral synthesis.

Theorem 4.16. *The function $f : \mathbb{Z} \times \mathbb{Z} \to \mathbb{C}$ satisfies the functional equation (4.9) for all x, y, t in $\mathbb{Z}$ if and only if*

$$f(x, y) = \varphi(x) + \psi(y) + \chi(x - y) \tag{4.21}$$

holds for all x, y in $\mathbb{Z}$, where $\varphi, \psi, \chi : \mathbb{Z} \to \mathbb{C}$ are arbitrary functions.

Proof. The "if" part is obvious. To prove the converse we observe that the solution space of (4.9) is obviously a translation invariant linear space which is closed with respect to pointwise convergence. By Theorem 2.22 of Lefranc on spectral synthesis for $\mathbb{Z}^n$, the linear hull of all exponential monomial solutions of (4.9) is dense in the solution space of (4.9). Therefore it is enough to show that any exponential monomial solution of (4.9) has the form (4.21), and that this form is preserved while taking pointwise limits. In what follows we will show this.

First, we determine all exponential solutions of (4.9). As complex exponential functions on $\mathbb{Z} \times \mathbb{Z}$ have the form $(x, y) \mapsto \lambda^x \mu^y$ with some nonzero complex numbers λ, μ, we substitute this function into (4.9). Then we have

$$(\lambda\mu - 1)(\lambda - 1)(\mu - 1) = 0,$$

which implies that we have the following cases:

1. $\mu = 1$, $\lambda \neq 1$ is arbitrary,
2. $\lambda = 1$, $\mu \neq 1$ is arbitrary,
3. $\mu = \lambda^{-1} \neq 1$ is arbitrary,
4. $\lambda = \mu = 1$.

Now we find the exponential monomial solutions of (4.9) in all these cases. In Case 1 suppose that the function $(x, y) \mapsto p(x, y)\lambda^x$ is a solution of (4.9), where $p : \mathbb{R} \times \mathbb{R} \to \mathbb{C}$ is a polynomial and $\lambda \neq 1$ is a nonzero complex number. Substitution gives

$$\begin{aligned} p(x + t, y + t)\lambda^t &+ p(x - t, y)\lambda^{-t} + p(x, y - t) \tag{4.22} \\ &= p(x - t, y - t)\lambda^{-t} + p(x, y + t) + p(x + t, y)\lambda^t \end{aligned}$$

for all x, y, t in $\mathbb{Z}$. As different exponential functions are linearly independent over the ring of polynomials (see [66], Lemma 4.3), we have that

$$p(x, y - t) = p(x, y + t)$$

holds for all x, y, t in $\mathbb{Z}$, which means that p is independent of the second variable, hence in this case any exponential monomial solution of (4.9) has the form

$$\varphi(x, y) = a(x)$$

for all x, y in $\mathbb{Z}$, where $a : \mathbb{Z} \to \mathbb{C}$ is a function. Similarly, in Case 2 we have that any exponential monomial solution of (4.9) has the form

$$\psi(x, y) = b(y)$$

for all x, y in $\mathbb{Z}$, where $b : \mathbb{Z} \to \mathbb{C}$ is a function. In Case 3 suppose that the function $(x, y) \mapsto p(x, y)\lambda^{x - y}$ is a solution of (4.9), where $p : \mathbb{R} \times \mathbb{R} \to \mathbb{C}$ is a polynomial, and $\lambda \neq 1$ is a nonzero complex number. Substitution gives

$$p(x + t, y + t) + p(x - t, y)\lambda^{-t} + p(x, y - t)\lambda^t \qquad (4.23)$$
$$= p(x - t, y - t) + p(x, y + t)\lambda^{-t} + p(x + t, y)\lambda^t$$

for all x, y, t in $\mathbb{Z}$. Again, as different exponential functions are linearly independent over the ring of polynomials, we have that

$$p(x + t, y + t) = p(x - t, y - t)$$

holds for all x, y, t in $\mathbb{Z}$. Hence by Theorem 4.14 it follows that in Case 3. any exponential monomial solution of (4.9) has the form $\chi(x, y) = c(x - y)$ for all x, y in $\mathbb{Z}$, where $c : \mathbb{R} \to \mathbb{C}$ is a polynomial.

Finally, in Case 4 suppose that the polynomial p is a solution of (4.9). Then by Theorem 4.15 we have that p has the form (4.15). Hence we infer that any exponential polynomial solution of (4.9) has the form (4.15). By spectral synthesis, any solution of (4.9) is the pointwise limit of exponential polynomial solutions, that is, functions of the form (4.21).

It is enough to show the following: If one has a pointwise convergent net of functions of the form

$$e(x, y) = \varphi(x) + \psi(y) + \chi(x + y)$$

(substitute $-y$ for y in the previous form, in order to get a symmetric expression), then the pointwise limit has the same form. Substituting $y = 0$ and then $x = 0$ we get

$$e(x, 0) = \varphi(x) + \psi(0) + \chi(x)$$

and

$$e(0, y) = \varphi(0) + \psi(y) + \chi(y),$$

respectively. Hence

$$e(x, y) - e(x, 0) - e(0, y) = \chi(x + y) - \chi(x) - \chi(y) - \varphi(0) - \psi(0)$$

which implies that the function

$$(x, y) \mapsto e(x, y) - e(x, 0) - e(0, y)$$

satisfies the *co-cycle equation*

$$F(x, y) + F(x + y, z) = F(x, y + z) + F(y, z).$$

It follows that the pointwise limit satisfies the same equation. By a result of M. Hosszú (see [25] and [26]) any solution of the co-cycle equation has the form

$$F(x, y) = \chi(x + y) - \chi(x) - \chi(y) + B(x, y),$$

where χ is arbitrary and B is bi-additive and antisymmetric. However, any antisymmetric bi-additive function on $\mathbb{Z} \times \mathbb{Z}$ is identically zero, hence we infer by spectral synthesis that

$$f(x, -y) = \varphi(x) + \psi(-y) + \chi(x + y)$$

and then

$$f(x, y) = \varphi(x) + \psi(y) + \chi(x - y).$$

Our theorem is now proved.

References: [25], [26], [60], [37], [66], [68], [52], [71].

5

Mean periodic functions

5.1 The Fourier transform of mean periodic functions

In Section 4.1 we have seen that the study of varieties in $\mathcal{C}(G)$ is equivalent to the study of solution spaces of convolution type functional equations. As the solutions of convolution type functional equations are mean periodic functions, it seems to be reasonable to set them into the center of our investigations. We recall (see also [10]) that for a locally compact Abelian group G the continuous function $f : G \to \mathbb{C}$ is called mean periodic if there exists a nonzero compactly supported complex Radon measure μ on G such that

$$f * \mu = 0.$$

For a given μ the set of all continuous functions f with this property will be denoted by $V(\mu)$. Obviously $V(\mu)$ is a variety, which is the annihilator of the ideal generated by μ^*. The union of all varieties of the form $V(\mu)$ with $\mu \neq 0$ endowed with the inductive limit of the topologies of the spaces $V(\mu)$ is a topological space, which we call the *space of mean periodic functions* and denote by $\mathcal{MP}(G)$. It is clear that the continuous function $f : G \to \mathbb{C}$ belongs to $\mathcal{MP}(G)$ if and only if $\tau(f) \neq \mathcal{C}(G)$. Convergence of a net in $\mathcal{MP}(G)$ means that this net is contained in the annihilator of some nonzero compactly supported complex Radon measure and it converges uniformly on compact sets.

If the varieties $V(\mu)$ for $\mu \neq 0$ in $\mathcal{M}_c(G)$ form an inductive set, that is, for any nonzero measures μ_1, μ_2 in $\mathcal{M}_c(G)$ there exists a nonzero measure μ_0 in $\mathcal{M}_c(G)$ with $V(\mu_1) \cup V(\mu_2) \subseteq V(\mu_0)$, then $\mathcal{MP}(G)$ is a locally convex topological vector space. This holds, for instance, if $\mu_1 * \mu_2 \neq 0$, that is, if the convolution of nonzero compactly supported measures is nonzero, which is obviously the case if $G = \mathbb{Z}$ or $G = \mathbb{R}$.

In the case $G = \mathbb{Z}$ a complex valued function $f : \mathbb{Z} \to \mathbb{C}$ is mean periodic if and only if it satisfies a nontrivial finite difference equation. From the theory of finite difference equations we know that this is the case if and only if f is an exponential polynomial, hence the theory of mean periodic functions on $\mathbb{Z}$ reduces to the theory of exponential polynomials and finite difference equations. We remark that the situation is completely different in $G = \mathbb{Z}^n$, where $n > 1$ is any integer, as in this case, for instance, all complex valued functions depending only on one variable are mean periodic.

Now we consider the case $G = \mathbb{R}$. In this case by the obvious isomorphism $\exp \lambda x \leftrightarrow \lambda$ for x in $\mathbb{R}$ and λ in $\mathbb{C}$ the set $\widetilde{\mathbb{R}}$ can be identified with $\mathbb{C}$. As we mentioned above, in this case $\mathcal{MP}(\mathbb{R})$ is a locally convex topological vector space. In particular, the sum of mean periodic functions is mean periodic, too. We also have the following theorem.

Theorem 5.1. *Any exponential polynomial is mean periodic on* $\mathbb{R}$.

Proof. It is enough to prove the statement for exponential monomials of the form $f = pm$, where $p : R \to \mathbb{C}$ is a polynomial of degree at most n and $m : \mathbb{R} \to \mathbb{C}$ is an exponential. Applying the difference operator

$$\Delta_y^{n+1} = \sum_{k=0}^{n+1} \binom{n+1}{k} (-1)^{n+1-k} \tau_y^k$$

to $p = f\breve{m}$ we have

$$\sum_{k=0}^{n+1} \binom{n+1}{k} (-1)^{n+1-k} f(x + ky)\breve{m}(y)^k = 0$$

for all real x, y. Obviously, this equation holds for any element g of $\tau(f)$ in place of f. Let m_0 be any exponential different from m and substitute m_0 for f into the left-hand side of the above equation with some real y for which $m_0(y) \neq m(y)$. We have

$$\sum_{k=0}^{n+1} \binom{n+1}{k} (-1)^{n+1-k} m_0(x + ky)\breve{m}(y)^k$$

$$= m_0(x) \sum_{k=0}^{n+1} \binom{n+1}{k} (-1)^{n+1-k} m_0(y)^k \breve{m}(y)^k$$

$$= m_0(x) \left(m_0(y)m(y)^{-1} - 1 \right)^{n+1},$$

which is different from zero for any real x. Hence m_0 is not in $\tau(f)$, and f is mean periodic.

We note that the same proof works for the same theorem in $\mathcal{C}(G)$ if all the varieties of the form $V(\mu)$ with nonzero μ in $\mathcal{M}_c(G)$ form an inductive family.

By the above theorem, on the set of reals all polynomials, exponential functions and continuous periodic functions are mean periodic. Hence $\widetilde{\mathbb{R}}$ is a subset of $\mathcal{MP}(\mathbb{R})$. However, the topology of $\mathcal{MP}(\mathbb{R})$ restricted to $\widetilde{\mathbb{R}}$ is discrete as is shown by the following theorem.

Theorem 5.2. *The set $\widetilde{\mathbb{R}}$ equipped with the topology of $\mathcal{MP}(\mathbb{R})$ is a discrete topological space.*

Proof. We show that any subset of $\widetilde{\mathbb{R}}$ is closed. Let $\{m_\alpha\}$ be a net of different exponentials on $\mathbb{R}$ which converges to the exponential m in the space $\mathcal{MP}(\mathbb{R})$. Let us suppose that $m_\alpha(x) = e^{\lambda_\alpha x}$ and $m(x) = e^{\lambda x}$ for all x in $\mathbb{R}$ with some λ_α and λ in $\mathbb{C}$. By the definition of the topology on $\mathcal{MP}(\mathbb{R})$ there exists a nonzero μ such that m_α is in $V(\mu)$, which implies that the Fourier–Laplace transform of μ vanishes at λ_α for any α. On the other hand, the net $\{e^{\lambda_\alpha x}\}$ converges to $\{e^{\lambda x}\}$ on any compact set in $\mathbb{R}$ uniformly, which implies that $\{\lambda_\alpha\}$ tends to λ. However, the Fourier–Laplace transform of μ is a nonzero entire function, and the zero set of a nonzero entire function cannot have a finite accumulation point, which is a contradiction, so our theorem is proved.

The product of a mean periodic function and an exponential is also mean periodic. Nevertheless, no compactly supported continuous function is mean periodic unless it is identically zero.

Another useful observation concerning derivatives is based on the identity

$$\left(pm * \mu\right)' = (pm)' * \mu\,,$$

which holds for any polynomial p, exponential m and measure μ in $\mathcal{M}_c(\mathbb{R})$. This implies that if n is a nonnegative integer, p is a polynomial of degree n, m is an exponential, μ is a measure in $\mathcal{M}_c(\mathbb{R})$ and the exponential monomial pm belongs to $V(\mu)$, then the functions $x \mapsto x^j m(x)$ for $j = 0, 1, \ldots, n$ all belong to $V(\mu)$. Indeed, the above identity shows that if pm belongs to $V(\mu)$, then $(pm)'$ belongs to $V(\mu)$, but $(pm)' = p'm + \lambda pm$, where $\lambda = m(0)$, hence $p'm$ belongs to $V(\mu)$. Continuing this process we have that the functions $pm, p'm, \ldots, p^{(n)}m$ all belong to $V(\mu)$, and obviously any function of the form $x \mapsto x^j m(x)$ with $j = 0, 1, \ldots, n$ is a linear combination of them. In particular, if the exponential monomial pm belongs to a variety, then, as we mentioned above, m itself belongs to it.

Based on the result of Schwartz on spectral synthesis in $\mathcal{C}(\mathbb{R})$ we introduce a generalized Fourier transform of mean periodic functions. To do so, we need a mean value operator on $\mathcal{MP}(\mathbb{R})$ (see [65], [66]).

Theorem 5.3. *There exists a unique continuous linear operator*

$$M : \mathcal{MP}(\mathbb{R}) \to \mathcal{P}(\mathbb{R})$$

with the properties

i) $M(\tau_y f) = \tau_y[M(f)]$,

ii) $M(p) = p$

for any mean periodic function f, polynomial p, and y in $\mathbb{R}$.

Proof. In the proof for any nonnegative integer k and exponential m we denote by φ_k the function $x \mapsto x^k m(x)$. First we prove the uniqueness. By Theorem 2.21 of Schwartz on spectral synthesis it is enough to show that the properties of M determine M on the set of all exponential monomials. We show that

$$M(\varphi_k) = 0$$

for any nonnegative integer k and for any exponential $m \neq 1$. By the properties of M we have $M(1) = 1$. Let $m \neq 1$ be an exponential. For any y in $\mathbb{R}$ it follows

$$M(m) = M[m(-y)\tau_y m] = m(-y)M(\tau_y m) = m(-y)\tau_y[M(m)],$$

and hence

$$M(m)m(y) = \tau_y[M(m)].$$

Here the function $y \mapsto \tau_y[M(m)(x)]$ is a polynomial for any x in $\mathbb{R}$. Thus, if $M(m)$ is not identically zero, then m is a polynomial, which is impossible unless $m = 1$, but this is excluded. Hence $M(m) = 0$ for any exponential $m \neq 1$. Suppose that we have proved $M[\varphi_j(y)] = 0$ for $j = 0, 1, \ldots, k - 1$. Then we have

$$M_y[(y + z)^k m(y + z)] = m(z)M_y\left[\sum_{j=0}^{k}\binom{k}{j}y^j z^{k-j}m(y)\right]$$

$$= m(z)\sum_{j=0}^{k}\binom{k}{j}z^{k-j}M_y[y^j m(y)] = m(z)M_y[y^k m(y)]$$

for any z in $\mathbb{R}$. (We write $M_y(f(y))$ in order to indicate that M is applied to the function in parentheses as a function of y.) In other words,

$$M[\tau_z \varphi_k] = m(z)M(\varphi_k)$$

holds for any z in $\mathbb{R}$. By the properties of M it follows that

$$\tau_z[M(\varphi_k)] = m(z)M(\varphi_k).$$

Similarly as above, we infer that the left-hand side is a polynomial in z, hence $M(\varphi_k) \neq 0$ would imply that m is a polynomial, which is impossible. Hence $M(\varphi_k) = 0$. This proves the uniqueness.

In order to prove the existence, first we notice that for any nonzero μ in $\mathcal{M}_c(\mathbb{R})$ the exponential 1 is not contained in the closed linear subspace of $\mathcal{C}(\mathbb{R})$ spanned by all exponential monomials in $V(\mu)$ different from 1. This follows from [55], Théoréme 7 and implies the existence of a measure μ_0 in $\mathcal{M}_c(\mathbb{R})$ such that $\langle 1, \mu_0^* \rangle = 1$ and $\langle \psi, \mu_0^* \rangle = 0$ for any exponential monomial $\psi \neq 1$ in $V(\mu)$. It follows that if $m \neq 1$ is an exponential and k is a nonnegative integer for which φ_k is in $V(\mu)$, then $\varphi_k * \mu_0 = 0$. Indeed, if φ_k is in $V(\mu)$, then φ_j is in $V(\mu)$, too, for $j = 0, 1, \ldots, k$, which means $\langle \varphi_j, \mu_0^* \rangle = 0$ for $j = 0, 1, \ldots, k$. On the other hand

$$\varphi_k * \mu_0(x) = \int \varphi_k(x - y)\, d\mu_0(y) = \int (x - y)^k m(x - y)\, d\mu_0(y)$$

$$= m(x) \int \sum_{j=0}^{k} \binom{k}{j} x^{k-j}(-y)^j m(-y)\, d\mu_0(y)$$

$$= m(x) \sum_{j=0}^{k} \binom{k}{j} x^{k-j} \int \varphi_j(-y)\, d\mu_0(y) = m(x) \sum_{j=0}^{k} \binom{k}{j} x^{k-j} \langle \varphi_j, \mu_0^* \rangle = 0.$$

However, if $m = 1$, then the above computation gives $\varphi_k * \mu_0 = \varphi_k$. This shows that $\varphi * \mu_0$ is a polynomial in $V(\mu)$ for any exponential polynomial φ in $V(\mu)$. But the degrees of all polynomials belonging to $V(\mu)$ have an upper bound, or else all polynomials would belong to $V(\mu)$ and the Stone–Weierstrass Theorem would imply $V(\mu) = \mathcal{C}(\mathbb{R})$, which is not the case. Let f be arbitrary in $V(\mu)$, then by spectral synthesis there exist exponential polynomials ψ_k in $V(\mu)$ with $\lim \psi_k = f$. Then we have $f * \mu_0 = \lim \psi_k * \mu_0$, hence also $f * \mu_0$ is a polynomial.

Suppose now that f also belongs to $V(\nu)$ with some nonzero ν in $\mathcal{M}_c(\mathbb{R})$. Then $f * \mu_0$ also belongs to $V(\nu)$, and it is a polynomial, therefore we have $f * \mu_0 = f * \mu_0 * \nu_0$. Similarly, $f * \nu_0 = f * \nu_0 * \mu_0$, which means that $f * \mu_0$ does not depend on the special choice of μ. On the other hand, any mean periodic function f is contained in some $V(\mu)$ with $\mu \neq 0$, hence we can define

$$M(f) = f * \mu_0$$

for any nonzero μ in $\mathcal{M}_c(\mathbb{R})$ with $f * \mu = 0$. The continuity and linearity of M follows from the definition, $i)$ follows from the properties of convolution and $ii)$ has been proved above.

In other words, the identity operator on the set of polynomials can be extended uniquely to a covariant continuous linear polynomial-valued operator on the set of all mean periodic functions. We call M the *mean value operator* on $\mathcal{MP}(\mathbb{R})$. We note that $\mathcal{P}(\mathbb{R})$ bears the topology inherited from $\mathcal{C}(\mathbb{R})$.

From the results of [33] it follows that the product of a mean periodic function and a polynomial is mean periodic, too. For this case we have the following theorem.

Theorem 5.4. *Let M denote the mean value operator on $\mathcal{MP}(\mathbb{R})$. Then we have*

$$M(pf) = pM(f)$$

for any polynomial p and mean periodic function f.

Proof. We have proved above that $M(pm) = 0$ if p is a polynomial and $m \neq 1$ is an exponential. This means that the statement of the theorem holds if f is an exponential monomial. By the theorem of Schwartz the linear combinations of exponential monomials are dense in $\mathcal{MP}(\mathbb{R})$, hence by the continuity of M the theorem is proved.

For a given exponential m and measure μ in $\mathcal{M}_c(\mathbb{R})$ we define the measure μ_m by the formula

$$\langle f, \mu_m \rangle = \langle f\check{m}, \mu \rangle$$

for any f in $\mathcal{C}(\mathbb{R})$. Obviously, μ_m is in $\mathcal{M}_c(\mathbb{R})$. If $f * \mu = 0$ holds for some f in $\mathcal{C}(\mathbb{R})$ and μ in $\mathcal{M}_c(\mathbb{R})$, then $f\check{m} * \mu_m = 0$. This means that for any mean periodic function f and for any exponential m the function $f\check{m}$ is also mean periodic. Hence we can define

$$\hat{f}(m) = M(f\check{m}).$$

Then $\hat{f} : \widetilde{\mathbb{R}} \to \mathcal{P}(\mathbb{R})$ is a polynomial-valued mapping on the set of all exponentials. It is called the *Fourier transform* of the mean periodic function f. By the isomorphism $\exp \lambda x \leftrightarrow \lambda$ between $\widetilde{\mathbb{R}}$ and $\mathbb{C}$ we can identify $\widetilde{\mathbb{R}}$ with $\mathbb{C}$ and $\hat{f}$ can be considered as a polynomial-valued function on $\mathbb{C}$. Although we can write $\hat{f}(\lambda)$ instead of $\hat{f}(m)$, sometimes we prefer not to use this identification.

We summarize some basic properties of this Fourier transform which follow directly from the definition.

Theorem 5.5. *The mapping $f \mapsto \hat{f}$ is linear and has the following properties:*

i) $(pf)\hat{} = p\hat{f}$,
ii) $(\tau_y f)\hat{}(m) = m(y)\tau_y[\hat{f}(m)]$

for any mean periodic function f, polynomial p, exponential m, and y in $\mathbb{R}$. Further, the mapping $f \mapsto \hat{f}(m)$ is continuous for any fixed exponential m.

We note that $\hat{1}(m) = 1$ for $m = 1$ and $\hat{1}(m) = 0$ for $m \neq 1$. Hence, for any polynomial p we infer $\hat{p}(m) = 0$ for $m \neq 1$ and $\hat{p}(1) = p$.

The following theorem expresses the injectivity of the Fourier transformation.

Theorem 5.6. *For any mean periodic function f the equation $\hat{f} = 0$ implies $f = 0$.*

Proof. If $\hat{f} = 0$, then the previous theorem implies that $\hat{\varphi} = 0$ for any φ in $\tau(f)$. In particular, $\hat{\varphi} = 0$ for any exponential polynomial φ in $\tau(f)$. Hence, by spectral synthesis it is enough to show that the theorem holds for exponential polynomials. Let f be an exponential polynomial of the form

$$f = \sum_{i=1}^{n} p_i m_i,$$

where $p_1, \ldots, p_n$ are polynomials and $m_1, m_2, \ldots, m_n$ are different exponentials. Then by the properties of the Fourier transformation we have $\hat{f}(m) = p_i$, if $m = m_i$ for some $i = 1, 2, \ldots, n$, and $\hat{f}(m) = 0$ if m is different from the m_i's. This implies our statement.

The following inversion theorem is also valid.

Theorem 5.7. *(Inversion Theorem) Let f be a mean periodic function and suppose that the series*

$$\sum_{m \in sp\, f} \hat{f}(m)m$$

is convergent in the Moore–Smith sense in $\mathcal{MP}(\mathbb{R})$. Then

$$f = \sum_{m \in sp\, f} \hat{f}(m)m.$$

Proof. The condition on the Moore–Smith convergence of the series guarantees that

$$[\sum_{m \in sp\, f} \hat{f}(m)m]\hat{} = \sum_{m \in sp\, f} [\hat{f}(m)m]\hat{},$$

hence, if

$$g = \sum_{m \in sp\, f} \hat{f}(m)m,$$

then g is mean periodic and we can easily check that $\hat{g} = \hat{f}$. Hence, by the previous theorem, $g = f$.

Theorem 5.8. *The mean periodic function f is an exponential polynomial if and only if $sp\, f$ is finite, and it is an exponential monomial if and only if $sp\, f$ is a singleton.*

This corollary can be considered as a generalization of the Primary Ideal Theorem.

Theorem 5.9. *For any mean periodic function f we have*

$$sp\, f = supp\, \hat{f}.$$

Proof. We remark that *supp* $\hat{f}$ is the set of all exponentials m for which the polynomial $\hat{f}(m)$ is not identically zero. First, we note that if φ is in $\tau(f)$, then $\tau(\varphi) \subseteq \tau(f)$, hence if the exponential m does not belong to $sp\,f$, then it does not belong to $sp\,\varphi$ for any exponential polynomial φ in $\tau(f)$. If the functions φ_n $(n = 1, 2, \dots)$ are exponential polynomials in $\tau(f)$ such that $\lim \varphi_n = f$, then $\lim \hat{\varphi}_n(m) = \hat{f}(m) = 0$, hence m does not belong to *supp* $\hat{f}$ and *supp* $\hat{f} \subseteq sp\,f$. To prove the reverse inclusion, suppose that m is not in *supp* $\hat{f}$, that is, $\hat{f}(m) = 0$. Then for any y in $\mathbb{R}$ we have

$$(\tau_y f)\hat{}(m) = m(y)\tau_y[\hat{f}(m)] = 0,$$

hence $\hat{g}(m) = 0$ for all g in $\tau(f)$. If n is an exponential in $sp\,f$, then n is in $\tau(f)$, hence $\hat{n}(m) = 0$, which implies $n \neq m$. This means that m is not in $sp\,f$, so the theorem is proved.

As a simple application we construct a spectral synthesis for a particular function. Let, for any x in $\mathbb{R}$,

$$\varphi(x) = \frac{1}{2} - (-1)^{[x]}\left(\frac{1}{2} - \{x\}\right).$$

This function is continuous and periodic by 2, hence it is mean periodic. Let, for any x in $\mathbb{R}$,

$$f(x) = x\varphi(x).$$

It is easy to check that f satisfies the difference equation

$$f(x+4) - 2f(x+2) + f(x) = 0$$

for all x in $\mathbb{R}$, hence it is mean periodic. We remark that f is neither exponential polynomial nor periodic.

To find the Fourier transform of f we first compute the Fourier series of the function φ. By easy computation we obtain the series

$$\frac{1}{2} - \frac{4}{\pi^2}\sum_{k=0}^{\infty}\frac{\cos(2k+1)\pi x}{(2k+1)^2},$$

which is obviously uniformly convergent. Then we have

$$f(x) = \frac{x}{2} - \frac{4x}{\pi^2}\sum_{k=0}^{\infty}\frac{\cos(2k+1)\pi x}{(2k+1)^2}$$

for any x in $\mathbb{R}$, and the series is uniformly convergent on any compact subset of $\mathbb{R}$. Identifying $\tilde{\mathbb{R}}$ with $\mathbb{C}$ as indicated above we have that

$$\hat{f}(0)(x) = \frac{x}{2},$$

and

$$\hat{f}\big(\pm (2k+1)\pi i\big)(x) = -\frac{2x}{(2k+1)^2\pi^2}$$

for any nonnegative integer k. For all other values of λ we have $\hat{f}(\lambda) = 0$.

5.2 The Fourier transform of exponential polynomials

In the previous section we have seen that the Fourier transform of mean periodic functions — due to its nice properties — may be used successfully to solve convolution type functional equations on the real line. In order to extend this transformation to mean periodic functions on more general Abelian groups, we have to assure the existence of the polynomial-valued covariant mean on the mean periodic functions, which depends heavily on spectral synthesis on $\mathcal{C}(\mathbb{R})$ and also on some special properties of $\mathbb{R}$. Although this extension is not possible in general, it can be defined at least on exponential polynomials on any locally compact Abelian group.

Let G be a locally compact Abelian group and let $\mathcal{EP}(G)$ denote the set of all exponential polynomials defined on G. We shall suppose that the varieties of the form $V(\mu)$, where μ is in $\mathcal{M}_c(G)$, form an inductive family, that is, for any nonzero measures μ_1, μ_2 in $\mathcal{M}_c(G)$ there exists a nonzero measure μ_0 in $\mathcal{M}_c(G)$ such that $V(\mu_1) \cup V(\mu_2) \subseteq V(\mu_0)$. As we mentioned above, this happens if the convolution of two nonzero measures in $\mathcal{M}_c(G)$ is nonzero. In this case $\mathcal{MP}(G)$ is a locally convex topological vector space and $\mathcal{EP}(G)$ is a (not necessarily) closed linear subspace of it. Similarly, the set $\mathcal{P}(G)$ of all polynomials on G is a (not necessarily) closed linear subspace of $\mathcal{EP}(G)$.

In order to define a covariant mean operator on $\mathcal{EP}(G)$, we need to know that different exponentials are independent over the ring of polynomials. This is the content of the following theorem (see [66]).

Theorem 5.10. *Let G be a locally compact Abelian group and k a positive integer. Suppose that $m_1, m_2, \ldots, m_k : G \to \mathbb{C}$ are different exponential functions, and $p_1, p_2, \ldots, p_k : G \to \mathbb{C}$ are polynomials. If the function $\sum_{i=1}^{k} p_i m_i$ vanishes on G, then $p_i = 0$ for $i = 1, 2, \ldots, k$.*

The above theorem implies that if $f : G \to \mathbb{C}$ is an exponential polynomial, then f can uniquely be represented in the form

$$f = p_0 + \sum_{i=1}^{k} p_i m_i,$$

where $m_1, m_2, \ldots, m_k : G \to \mathbb{C}$ are exponentials, different from 1 and from each other, and $p_0, p_1, \ldots, p_k : G \to \mathbb{C}$ are polynomials. We will call this the *canonical representation* of the exponential polynomial f. The polynomial p_0 in this representation will be denoted by $M(f)$. It is easy to check that the mapping $M : \mathcal{EP}(G) \to \mathcal{P}(G)$ enjoys the basic properties of the covariant mean of mean periodic functions as it is expressed in the following theorem.

Theorem 5.11. *The mapping* $M : \mathcal{EP}(G) \to \mathcal{P}(G)$ *defined above is linear and has the following properties:*

i) $M(\tau_y f) = \tau_y [M(f)]$,
ii) $M(p) = p$,
iii) $M(pf) = p M(f)$,
iv) $M(\check{f}) = [M(f)]\check{}$

for any exponential polynomial f, *polynomial* p, *and* y *in* G.

Linearity and properties $i)$, $ii)$ define M uniquely, hence in the real case the operator M defined above on mean periodic functions is a unique extension of this M. Using M we define $\hat{f}$ for any exponential polynomial $f : G \to \mathbb{C}$ by

$$\hat{f}(m) = M(f \cdot \check{m})$$

whenever $m : G \to \mathbb{C}$ is an exponential. Clearly, $f \cdot \check{m}$ is an exponential polynomial and we realize $\hat{f}(m)$ as the polynomial coefficient of m in the canonical representation of f. The function $\hat{f} : \widetilde{G} \to \mathcal{P}(G)$ is called the *Fourier transform* of the exponential polynomial f. Obviously, $\hat{f}$ is finitely supported and the values of f are constants if f is a trigonometric polynomial. In fact, in this case the support of $\hat{f}$ is contained in $\widehat{G}$ and $\hat{f}$ coincides with the ordinary Fourier transform of f.

The fundamental properties of the map $f \mapsto \hat{f}$ are summarized in the following theorem which can be proved easily by Theorem 5.11.

Theorem 5.12. *The map* $f \mapsto \hat{f}$ *defined above is linear and has the following properties:*

i) $\hat{p}(m) = 0$ *for* $m \neq 1$,
ii) $(pf)\hat{}(m) = p \cdot \hat{f}(m)$,
iii) $(\tau_y f)\hat{}(m) = m(y) \cdot (\tau_y \hat{f})(m)$,
iv) $(\check{f})\hat{}(m) = [\hat{f}(\check{m})]\check{}$

for all exponential polynomials f, *polynomials* p, *exponentials* m *and* y *in* G.

We also have the following "inversion theorem".

Theorem 5.13. *Let* $f : G \to \mathbb{C}$ *be an exponential polynomial. Then*

$$f = \sum_{m \in \widetilde{G}} \hat{f}(m) m .$$

5.3 Applications to differential equations

In the case $G = \mathbb{R}^n$ we can describe easily the behavior of the Fourier transformation of exponential polynomials with respect to differentiation which is useful for the applications to ordinary and partial differential equations. As any exponential m on $\mathbb{R}^n$ has the form $x \mapsto \exp\langle\lambda, x\rangle$ with some λ in $\mathbb{C}^n$, we identify m with λ and we write $\hat{f}(\lambda)$ instead of $\hat{f}(m)$. We denote by ∂ the vector-valued operator $(\partial_1, \partial_2, \ldots, \partial_n)$, where ∂_i is the usual operator of partial differentiation with respect to the i-th variable for $i = 1, 2, \ldots, n$. In the case $n = 1$ we write D instead of ∂. If α is any n-dimensional multi-index, then

$$\partial^\alpha = \partial_1^{\alpha_1} \partial_2^{\alpha_2} \ldots \partial_n^{\alpha_n}.$$

If λ is in $\mathbb{C}^n$, then $\partial + \lambda$ denotes the vector-valued operator $\partial + \lambda \cdot I$, where I denotes the identity operator. Substituting the n-dimensional vector ξ into the complex polynomial P in n variables has its obvious usual meaning. We also write $\langle\lambda, \partial\rangle$ for the operator $\lambda_1\partial_1 + \lambda_2\partial_2 + \cdots + \lambda_n\partial_n$.

Theorem 5.14. *Let us suppose that P is a complex polynomial in n variables and $f : \mathbb{R}^n \to \mathbb{C}$ is an exponential polynomial. Then for any λ in $\mathbb{C}^n$ we have*

$$(P(\partial)f)\hat{}(\lambda) = P(\partial + \lambda)\hat{f}(\lambda).$$

Proof. Clearly, the operator M introduced above commutes with partial differentiation. Let $m(x) = \exp\langle\lambda, x\rangle$ for any x in $\mathbb{R}^n$ and λ in $\mathbb{C}^n$, then for any $j = 1, 2, \ldots, n$ we have

$$(\partial_j f)\hat{}(\lambda) = M(\partial_j f \cdot \check{m}) = M[\partial_j(f \cdot \check{m}) - f \cdot \partial_j \check{m}]$$

$$= M[\partial_j(f \cdot \check{m})] - M(f \cdot \partial_j \check{m}) = \partial_j M(f \cdot \check{m}) - M[f \cdot (-\lambda_j)\check{m}]$$

$$= \partial_j \hat{f}(\lambda) + \lambda_j \cdot \hat{f}(\lambda) = (\partial_j + \lambda_j)\hat{f}(\lambda).$$

Repeating this argument we have our statement.

In the case $n = 1$ we have the following result.

Theorem 5.15. *Let $f : \mathbb{R} \to \mathbb{C}$ be an exponential polynomial and k a nonnegative integer. Then for any complex λ we have*

$$(D^k f)\hat{}(\lambda) = (D + \lambda)^k \hat{f}(\lambda).$$

As a simple application we show how to find all solutions of inhomogeneous linear differential equations of the form

$$P(D)y = f \tag{5.1}$$

where P is a complex polynomial of degree n and f is a given exponential polynomial. It turns out that (5.1) always has an exponential polynomial solution.
As all solutions of the corresponding homogeneous equation are exponential
polynomials, it follows that all solutions of (5.1) are exponential polynomials.
Let y denote any solution of (5.1). Applying Fourier transformation on both
sides of (5.1) we have that

$$P(D + \lambda)\hat{y}(\lambda) = \hat{f}(\lambda)$$

holds for each complex λ. Here $\hat{y}(\lambda)$ and $\hat{f}(\lambda)$ are polynomials and $\hat{f}(\lambda) = 0$
except for finitely many values of λ. Hence, the problem of solving (5.1) is
reduced to the problem of finding all polynomial solutions q of an equation of
the form

$$P(D + \lambda)q = p,$$

where p is a given polynomial. It is easy to derive simple systems of linear
equations for the coefficients of the polynomial q which makes it possible to
find all solutions of (5.1) without integration. For the details see [63]. We
remark that the same ideas can be applied to find all exponential polynomial solutions of inhomogeneous linear differential equations with polynomial
coefficients.

In the case $n > 1$ the Fourier transform of exponential polynomials can
be applied to find exponential polynomial solutions of inhomogeneous linear
partial differential equations with constant or polynomial coefficients if the
inhomogeneous term is an exponential polynomial. Its application will be il-
lustrated on the Cauchy problem for the heat equation, that is on the problem

$$\partial_t u = a^2 \Delta_x u, \qquad (5.2)$$
$$u(x, 0) = u_0(x),$$

where $u_0 : \mathbb{R}^n \to \mathbb{C}$ is a given exponential polynomial. We try to find an
exponential polynomial $u : \mathbb{R}^n \times \mathbb{R} \to \mathbb{C}$, which is a solution of (5.2). Here ∂_t
denotes the partial differential operator ∂_{n+1} on $\mathbb{R}^n \times \mathbb{R}$ and $\Delta_x = \sum_{i=1}^n \partial_i^2$
is the Laplacian on $\mathbb{R}^n$.

Observing that all exponentials m on $\mathbb{R}^n \times \mathbb{R}$ have the form

$$m(x, t) = \exp(\langle \lambda, x \rangle + \mu t)$$

with some λ in $\mathbb{C}^n$ and μ in $\mathbb{C}$, the value of the Fourier transform of u at the
exponential m will be denoted by $\hat{u}(\lambda, \mu)$. Applying Fourier transformation
on both sides of the first equation of (5.2) we have

$$\partial_t \hat{u}(\lambda, \mu) + \mu \cdot \hat{u}(\lambda, \mu) = a^2 (\Delta_x + 2\langle \lambda, \partial \rangle + \langle \lambda, \lambda \rangle) \, \hat{u}(\lambda, \mu). \qquad (5.3)$$

Here $\langle \lambda, \partial \rangle = \sum_{i=1}^n \lambda_i \partial_i$. We know that $\hat{u}(\lambda, \mu)$ is a polynomial in (x, t) for
all λ in $\mathbb{C}^n$ and μ in $\mathbb{C}$. For a fixed pair λ, μ let

$$\hat{u}(\lambda, \mu)(x, t) = a_N(x)t^N + a_{N-1}(x)t^{N-1} + \cdots + a_0(x),$$

where a_k is a polynomial for $k = 0, 1, \ldots, N$ and $a_N \neq 0$. Substituting into (5.3) and comparing the coefficients of t^N we have $\mu = a^2\langle\lambda, \lambda\rangle$. Then comparing the coefficients of t^k for $k = 0, 1, \ldots, N - 1$ we have

$$a_{k+1}(x) = \frac{a^2}{k+1}\left(\Delta_x + 2\langle\lambda, \partial\rangle\right)a_k(x)$$

for all x in $\mathbb{R}^n$. Obviously, $a_0(x) = \hat{u}_0(\lambda)(x)$, hence

$$a_k(x) = \frac{a^2}{k!}\left(\Delta_x + 2\langle\lambda, \partial\rangle\right)^k \hat{u}_0(\lambda)(x)$$

holds for $k = 0, 1, \ldots, N$. Here N denotes the smallest nonnegative integer for which $a_N \neq 0$ and $a_{N+1} = 0$. The existence of such N follows from the fact that $\hat{u}_0(\lambda)$ is a polynomial, which is supposed to be nonzero.

Using the inversion formula we have

Theorem 5.16. *Let $u_0 : \mathbb{R}^n \to \mathbb{C}$ be an exponential polynomial. Then the unique solution of the Cauchy problem* (5.2) *is given by*

$$u(x, t) = \sum_{\lambda \in \mathbb{C}^n} \sum_{N=0}^{\infty} \frac{[a^2(\Delta_x + 2\langle\lambda, \partial\rangle)]^N}{N!} \, \hat{u}_0(\lambda)(x) \cdot t^N \cdot e^{\langle\lambda, x\rangle + a^2\langle\lambda, \lambda\rangle t}$$

for all x in $\mathbb{R}^n$ and t in $\mathbb{R}$.

Here both sums are actually finite. A straightforward extension of this result can be obtained for the inhomogeneous Cauchy problem

$$\partial_t u = a^2 \Delta_x u + f(x, t),$$
$$u(x, 0) = u_0(x),$$

where $u_0 : \mathbb{R}^n \to \mathbb{C}$ and $f : \mathbb{R}^n \times \mathbb{R} \to \mathbb{C}$ are exponential polynomials. The ideas extend easily also to Cauchy problems for the Schrödinger, bi-harmonic or other equations if the given functions are exponential polynomials. For the details see [64].

References: [10], [55], [33], [63], [64], [65], [66].

6

Difference equations in several variables

6.1 Spectral synthesis of difference equations

The result of Lefranc on spectral synthesis in $F(\mathbb{Z}^n)$ can be used to give a simple method for the solution of linear systems of homogeneous difference equations with constant coefficients. The method is based on the simple fact that varieties in $F(\mathbb{Z}^n)$ are exactly the solution spaces of such systems of equations. In the case $n = 1$ the situation reduces to the classical theory of linear homogeneous difference equations with constant coefficients, as it has been exhibited in Section 2.4. Now we present a more detailed analysis of this subject in several variables (see [74]). First we recall and adjust our previous notation to the present situation.

Let k be a fixed positive integer. If $\lambda = (\lambda_1, \lambda_2, \ldots, \lambda_k)$ is an element of $\mathbb{C}^k$ with nonzero components and $x = (x_1, x_2, \ldots, x_k)$ is in $\mathbb{Z}^k$, then λ^x is defined as the product $\lambda_1^{x_1} \lambda_2^{x_2} \ldots \lambda_k^{x_k}$. Also, for any $i = 1, 2, \ldots, k$ we keep on using the former notation e_i for the vector whose i-th component is 1 and all the others are 0.

The *partial translation operators* τ_i are defined for $i = 1, 2, \ldots, k$ on functions $f : \mathbb{Z}^k \to \mathbb{C}$ by

$$\tau_i f(x) = f(x + e_i),$$

where x is in $\mathbb{Z}^k$. If we write symbolically $\tau = (\tau_1, \tau_2, \ldots, \tau_k)$, then — in accordance with the above notation — we can write for each $y = (y_1, y_2, \ldots, y_k)$ in $\mathbb{Z}^k$,

$$\tau^y = \tau_1^{y_1} \tau_2^{y_2} \ldots \tau_k^{y_k},$$

and for any $f : \mathbb{Z}^k \to \mathbb{C}$ we have

$$f(x + y) = \tau^y f(x)$$

whenever x, y are in $\mathbb{Z}^k$. Here τ^y is the *translation operator with increment* y. In particular, τ^0 is the identity operator.

Similar notation will be used for partial differential operators, acting on complex polynomials in k variables, as in the previous sections. In particular, for any y in $\mathbb{Z}^k$ we shall use the notation $\langle y, \partial \rangle$ for the differential operator $y_1 \partial_1 + y_2 \partial_2 + \cdots + y_k \partial_k$. Hence, for each nonnegative integer j the operator $\langle y, \partial \rangle^j$ is the j-th *differential*, which acts on the polynomial $p : \mathbb{Z}^k \to \mathbb{C}$ in the obvious manner. By the Taylor Formula we have for any polynomial $p : \mathbb{Z}^k \to \mathbb{C}$ and for any x, y in $\mathbb{Z}^k$,

$$p(x + y) = \sum_{j=0}^{\infty} \frac{1}{j!} \langle y, \partial \rangle^j p(x).$$

As p is a polynomial, this is a finite sum.

If λ is in $\mathbb{C}^k$ with nonzero components and $p : \mathbb{Z}^k \to \mathbb{C}$ is a polynomial, then the function $x \mapsto p(x)\lambda^x$ on $\mathbb{Z}^k$ is a typical exponential monomial and a linear combination of exponential monomials is an exponential polynomial.

We have seen above that varieties in $F(\mathbb{Z}^k)$ have a particularly simple form. Indeed, if Γ is a nonempty set and $c_\gamma : \mathbb{Z}^k \to \mathbb{C}$ is a finitely supported function for any γ in Γ, then all solutions $f : \mathbb{Z}^k \to \mathbb{C}$ of the system of linear difference equations

$$\sum_{y \in \mathbb{Z}^k} c_\gamma(y) f(x + y) = 0, \tag{6.1}$$

where x is in $\mathbb{Z}^k$ and γ is in Γ, form a variety, which is proper if and only if at least one of the functions c_γ is nonidentically zero. This variety is equal to the intersection of the kernels of all operators $\sum_{y \in \mathbb{Z}^k} c_\gamma(y) \tau^y$ with γ in Γ. Hence, any system of linear difference equations of the form (6.1) generates a variety, which is the solution space of the system. Conversely, obviously any variety arises in this way.

In the language of difference equations the spectral synthesis result, Theorem 2.22 of Lefranc, means that the exponential monomial solutions of a system of linear difference equations generate a dense set in the solution space. In other words, any solution is the pointwise limit of a sequence of exponential polynomial solutions. Hence, the exponential monomial solutions of a system of the form (6.1) characterize the whole solution space, and it seems to be reasonable to find methods for the determination of exponential monomial solutions.

The set of all exponential solutions of a system of the form (6.1) we called the spectrum of the system. We recall that the exponentials can be identified by elements of $\mathbb{C}^k$ with nonzero components. The following result is a reformulation of a well known fact, and it is straightforward to prove.

Theorem 6.1. *The spectrum of* (6.1) *is the set of all solutions* λ *in* $\mathbb{C}^k$ *with nonzero components of the system of algebraic equations*

$$\sum_{y \in \mathbb{Z}^k} c_\gamma(y) \lambda^y = 0 \qquad (\gamma \in \Gamma). \tag{6.2}$$

The left-hand sides of (6.2) are polynomials in the components of λ and their reciprocals, for any fixed γ in Γ. As the functions c_γ vanish off a finite set (depending on γ), each equation of the system (6.2) can be multiplied by an appropriate power of λ so that the new equations contain only nonnegative powers of the components of λ. As the new system, obtained in this way, is obviously equivalent to the original one, we may always suppose that in (6.2) $c_\gamma(y) = 0$ if y has a negative component. Then the left-hand sides of (6.2) are polynomials in y. The system (6.2) of algebraic equations is called the *system of characteristic equations* of (6.1), and the left-hand sides are the *characteristic polynomials* of (6.2). If $k = 1$ and Γ is a singleton, that is, if we have the case of a single linear homogeneous difference equation in one variable, then the system of characteristic equations reduces to a single equation, to the characteristic equation, corresponding to the characteristic polynomial. The nonzero roots of this polynomial form the spectrum of the equation. The set of all exponential monomial solutions of a system of the form (6.1) is the spectral set of the system. Hence, the spectral set can be identified with a set of pairs (λ, p), where λ is in $\mathbb{C}^k$ with nonzero components, and p is a polynomial for which the exponential monomial $x \mapsto \lambda^x p(x)$ is a solution of (6.1). The following result shows that the elements (λ, p) of the spectral set are "built up" from the spectrum. We have used this theorem earlier in the form that if an exponential monomial belongs to a variety, then the corresponding exponential belongs to it, too.

Theorem 6.2. *If* (λ, p) *is in the spectral set of* (6.1) *and* p *is a nonzero polynomial, then* λ *is in the spectrum of* (6.1).

Proof. Substituting the exponential monomial $x \mapsto \lambda^x p(x)$ into (6.1) we obtain

$$\sum_{y \in \mathbb{Z}^k} c_\gamma(y) \lambda^x \lambda^y p(x+y) = \sum_{y \in \mathbb{Z}^k} c_\gamma(y) \lambda^{x+y} p(x+y) = 0,$$

and hence

$$\sum_{y \in \mathbb{Z}^k} c_\gamma(y) \lambda^y p(x+y) = 0 \tag{6.3}$$

holds for any x in $\mathbb{Z}^k$ and γ in Γ. Here the left-hand side is a polynomial in x and p is nonzero, hence comparing the leading terms on both sides we get our statement.

This theorem shows that the determination of the spectral set of a system of the form (6.1) should start with the determination of the spectrum. This requires us to find the common roots of a set of given polynomials in k variables. However, this is only a part of the work to be done: we have to find all exponential monomials corresponding to these roots, which are solutions. This leads to finding polynomial solutions of the system (6.3). We show that this problem can be reduced to find polynomial solutions of systems of linear homogeneous partial differential equations.

For a given λ in $\mathbb{C}^k$ with nonzero coefficients the set of all polynomials $p : \mathbb{Z}^k \to \mathbb{C}$ satisfying (6.3) for any x in $\mathbb{Z}^k$ and γ in Γ is a translation invariant linear space of polynomials. It is not necessarily closed. A polynomial $p : \mathbb{Z}^k \to \mathbb{C}$ is a solution of (6.3) if and only if the exponential monomial $x \mapsto \lambda^x p(x)$ belongs to the spectral set of (6.1). The set of all linear operators of the polynomial ring $\mathbb{C}[z_1, z_2, \ldots, z_k]$ which are zero on the polynomial solutions of (6.3) is obviously an ideal. We call this ideal the *annihilator ideal* of (6.3). We present a generating set of this annihilator ideal consisting of homogeneous linear partial differential operators. Then the polynomial solutions of the system of difference equations (6.3) can be found by solving systems of linear partial differential equations in the ring of polynomials. The basic observation is the following.

Theorem 6.3. *The annihilator ideal of* (6.3) *is generated by the differential operators*

$$\sum_{y \in \mathbb{Z}^k} c_\gamma(y) \lambda^y \langle y, \partial \rangle^j \tag{6.4}$$

for all γ in Γ and for $j = 0, 1, \ldots$.

Proof. First of all we note that a polynomial in k variables with complex coefficients is a solution of (6.3) on $\mathbb{Z}^k$ if and only if it is a solution of (6.3) on $\mathbb{R}^k$. Hence we have to show that a polynomial $p : \mathbb{Z}^k \to \mathbb{C}$ is a solution of (6.3) if and only if

$$\sum_{y \in \mathbb{Z}^k} c_\gamma(y) \lambda^y \langle y, \partial \rangle^j p(x) = 0 \tag{6.5}$$

holds for all x in $\mathbb{R}^k$, for all γ in Γ and for $j = 0, 1, \ldots$. By the Taylor Formula we have

$$p(x + y) = \sum_{j=0}^{\infty} \frac{1}{j!} \langle y, \partial \rangle^j p(x)$$

for all x, y in $\mathbb{R}^k$. Obviously, the sum is finite. Substituting into (6.3) we have

$$\sum_{j=0}^{\infty} \frac{1}{j!} \sum_{y \in \mathbb{Z}^k} c_\gamma(y) \lambda^y \langle y, \partial \rangle^j p(x) = 0 \tag{6.6}$$

for any x in $\mathbb{R}^k$ and γ in Γ. On the left-hand side the j-th term is either zero or it is a homogeneous polynomial of degree exactly j. Hence (6.6) is equivalent to (6.5) for all x in $\mathbb{R}^k$, for all γ in Γ and for $j = 0, 1, \ldots$ and our theorem is proved.

If we use the convention $0^0 = 1$, then the system (6.5) includes the system of characteristic equations of (6.2) for $j = 0$. We can call the system of partial differential equations (6.5) the *characteristic differential equation system* of the system of difference equations (6.1). In this system λ and p are the unknowns and the pairs (λ, p) of solutions characterize the spectral set of (6.1). We note that λ has nonzero coefficients. If we have the spectral set of (6.1), then by Theorem 2.22 we know that any solution of (6.1) is the pointwise limit of linear combinations of spectral elements. So any property enjoyed by all the elements of the spectral set which is preserved under taking linear combinations and pointwise limits is possessed by any solution. Depending on the particular form of the equations (6.1) this may lead to the complete description of all solutions. We exhibit some particular examples for the application of this method in the following section.

6.2 Applications

Example 1. In this example we show how this method relates to the well-known solution method of linear difference equations in one variable. We let $k = 1$ and consider the difference equation

$$\sum_{l=0}^{n} c_l f(x + l) = 0, \tag{6.7}$$

supposing that n is a nonnegative integer, c_l is a complex number for $l = 0, 1, \ldots, n$ and $c_0 \neq 0$, $c_n \neq 0$.

The characteristic differential equation system of (6.7) now reduces to the system of differential equations

$$\sum_{l=0}^{n} c_l \lambda^l l^j p^{(j)}(x) = 0 \tag{6.8}$$

for $j = 0, 1, \ldots$ and x in $\mathbb{R}$. If $l^{(j)} = \frac{l!}{j!}$, then the system (6.8) is equivalent to the system

$$P^{(j)}(\lambda) \cdot p^{(j)}(x) = \sum_{l=0}^{n} c_l \lambda^{l-j} l^{(j)} p^{(j)}(x) = 0 \tag{6.9}$$

for $j = 0, 1, \ldots$ and x in $\mathbb{R}$, where

$$P(\lambda) = \sum_{l=0}^{n} c_l \lambda^l$$

is the characteristic polynomial of (6.7). By Theorem 6.3 we have that any exponential monomial solution f of (6.7) has the form

$$f(x) = \sum_{i=1}^{s} \lambda_i^x p_i(x) \tag{6.10}$$

for any x in $\mathbb{R}$, where the complex numbers $\lambda_1, \lambda_2, \ldots, \lambda_s$ are the different roots of the characteristic polynomial P with multiplicities $n_1, n_2, \ldots, n_s$, and p_i is a polynomial of degree $n_i - 1$ ($i = 1, 2, \ldots, s$). As the linear space of all functions of the form (6.10) is of finite dimension, it is a closed subspace in $\mathcal{C}(\mathbb{Z})$. Then Theorem 2.22 implies that any solution of (6.7) has the form (6.10). Thus from Theorem 6.3 one can derive the classical results for linear homogeneous difference equations with constant coefficients.

Example 2. We consider the difference equation

$$\sum_{l=0}^{n+1} \binom{n+1}{l} (-1)^{n+1-l} f(x+ly) = 0 \tag{6.11}$$

for all x, y in $\mathbb{Z}^k$, where n is a fixed nonnegative integer. Equation (6.11) can be written in the form

$$(\tau^y - I)^{n+1} f(x) = 0$$

for all x, y in $\mathbb{Z}^k$. Here I stands for the identity operator. The characteristic differential equation system of (6.11) has the form

$$\sum_{l=0}^{n+1} \binom{n+1}{l} (-1)^{n+1-l} \lambda^{ly} l^j \langle y, \partial \rangle^j p(x) = 0 \tag{6.12}$$

for $j = 0, 1, \ldots$ and x in $\mathbb{R}^k$. If $j = 0$, then we have

$$\left(\sum_{l=0}^{n+1} \binom{n+1}{l} (-1)^{n+1-l} (\lambda^y)^l \right) p(x) = 0$$

for all x in $\mathbb{R}^k$ and y in $\mathbb{Z}^k$, hence there are nonzero solutions if and only if

$$(\lambda^y - 1)^{n+1} = \sum_{l=0}^{n+1} \binom{n+1}{l} (-1)^{n+1-l} (\lambda^y)^l = 0$$

holds for each y in $\mathbb{Z}^k$. This implies $\lambda_1 = \lambda_2 = \cdots = \lambda_k = 1$, hence the spectral set of (6.11) consists of polynomials. Then (6.12) takes the form

$$\left(\sum_{l=0}^{n+1} \binom{n+1}{l} (-1)^{n+1-l} l^j \right) \langle y, \partial \rangle^j p(x) = 0$$

for all x in $\mathbb{R}^k$ and y in $\mathbb{Z}^k$ and for $j = 0, 1, \ldots$. The sum between the brackets is zero for $j \leq n$, and is equal to $(n+1)!$ for $j = n+1$, hence the differential operators $\langle y, \partial \rangle^{n+1}$ annihilate any polynomial solution of (6.11) for any y in $\mathbb{Z}^k$. In other words, the differential $\langle y, \partial \rangle^{n+1} p$ is zero for any y in $\mathbb{Z}^k$, hence the spectral set of (6.11) consists of polynomials of degree at most n. Thus, by Theorem 2.22 the solution space of (6.11) is the set of all polynomials of degree at most n (see [66]).

Example 3. We consider the difference equation

$$f(x+2, y) - 2f(x+1, y+1) + f(x, y+2) = 0 \tag{6.13}$$

on $\mathbb{Z}^2$. The characteristic differential equation system of (6.13) has the form

$$\left[\lambda^2 (2 \cdot \partial_1 + 0 \cdot \partial_2)^j - 2\lambda\mu(1 \cdot \partial_1 + 1 \cdot \partial_2)^j + \mu^2 (0 \cdot \partial_1 + 2 \cdot \partial_2)^j \right] p(x, y) = 0$$

for all x, y in $\mathbb{R}$ and $j = 0, 1, \ldots$. For $j = 0$ we have $\lambda^2 - 2\lambda\mu + \mu^2 = 0$ which implies $\lambda = \mu$. Using this it follows for $j = 1, 2 \ldots$

$$(2^j \partial_1^j - 2(\partial_1 + \partial_2)^j + 2^j \partial_2^j) p(x, y) = 0$$

for all x, y in $\mathbb{R}$. If $j = 1$, then we have no restriction on p, but for $j = 2$ we have the partial differential equation

$$(\partial_1 - \partial_2)^2 p(x, y) = 0$$

for all x, y in $\mathbb{R}$. If $q = (\partial_1 - \partial_2)p$, then $(\partial_1 - \partial_2)q = 0$, and this means that $q(x, y) = a(x+y)$ for all x, y in $\mathbb{R}$, where $a : \mathbb{R} \to \mathbb{C}$ is an arbitrary polynomial. Then again we have that

$$(\partial_1 - \partial_2)(p(x, y) - xa(x+y)) = 0,$$

hence $p(x, y) = xa(x+y) + b(x+y)$ for all x, y in $\mathbb{R}$, where $b : \mathbb{R} \to \mathbb{C}$ is an arbitrary polynomial, too. Hence, the spectral set of (6.13) consists of functions of the form

$$(x, y) \mapsto (xa(x+y) + b(x+y))\lambda^{x+y},$$

where λ is any nonzero complex number, and $a, b : \mathbb{R} \to \mathbb{C}$ are arbitrary polynomials. In particular, any function φ in the spectral set has the form

$$\varphi(x, y) = xA(x+y) + B(x+y) \tag{6.14}$$

for all x, y in $\mathbb{R}$, with some functions $A, B : \mathbb{R} \to \mathbb{C}$. As any such function also satisfies

$$\varphi(x, y) = x[\varphi(1, x + y - 1) + \varphi(0, x + y)] + \varphi(0, x + y)$$

for all x, y in $\mathbb{Z}$, this latter equation is also satisfied by pointwise limits of linear combinations of such functions. This means that any solution of (6.13) has the form (6.14). On the other hand, any function of the form (6.14) satisfies (6.13), hence the general solution of (6.13) is

$$f(x, y) = xA(x + y) + B(x + y)$$

with arbitrary functions $A, B : \mathbb{Z} \to \mathbb{C}$.

Example 4. Now we consider in three dimensions the difference equation

$$f(x + 2, y, z) - 2f(x + 1, y + 1, z + 1) = 0. \tag{6.15}$$

Here the characteristic differential equation system has the form

$$\left(\lambda^2 (2\partial_1)^j - 2\lambda\mu\nu(\partial_1 + \partial_2 + \partial_3)^j \right) p(x, y, z) = 0 \tag{6.16}$$

for all x, y, z in $\mathbb{R}$. For $j = 0$ we have $\lambda = 2\mu\nu$, which means that the exponential solutions of (6.15) have the form

$$(x, y, z) \mapsto 2^x \mu^{x+y} \nu^{x+z}$$

with some nonzero complex numbers μ, ν. This form may suggest that the general form of the solutions is

$$(x, y, z) \mapsto 2^x a(x + y, x + z) \tag{6.17}$$

with an arbitrary function $a : \mathbb{Z}^2 \to \mathbb{C}$, which are solutions, indeed. This suggestion is reinforced by the solutions of the characteristic differential equation system, which has the form

$$(2^j \partial_1^j - (\partial_1 + \partial_2 + \partial_3)^j) p(x, y, z) = 0$$

for all x, y, z in $\mathbb{R}$. Namely, for $j = 1$ we have the linear homogeneous partial differential equation

$$(\partial_1 - \partial_2 - \partial_3) p(x, y, z) = 0,$$

which implies that $p(x, y, z) = q(x + y, x + z)$ for all x, y, z with some polynomial $q : \mathbb{R}^2 \to \mathbb{C}$. This means that the spectral set of (6.15) consists of functions of the form (6.17), and as this form is preserved under taking linear combinations and pointwise limits, functions of the form (6.17) represent all solutions of (6.15), where $a : \mathbb{Z}^2 \to \mathbb{C}$ is an arbitrary function.

References: [66], [9], [74].

7

Spectral analysis and synthesis on polynomial hypergroups in a single variable

7.1 Polynomial hypergroups in one variable

In this section we formulate the basic problems of spectral analysis and spectral synthesis on commutative hypergroups and solve these problems on some types of hypergroups (see [72]). For more about L^1-spectral synthesis on hypergroups we refer to the the paper [78]. In [7] a Wiener Tauberian Theorem is presented for commutative locally compact hypergroups, whose dual is a hypergroup under pointwise operations. For further references on L^1-spectral synthesis in hypergroups the reader is referred to [8], [31], [50].

The concept of a DJS hypergroup (referring to the initials of C. F. Dunkl, R. I. Jewett and R. Spector) depends on a set of axioms which can be formulated in several different ways. Here we follow R. Lasser (see e.g. [51]). One begins with a locally compact Haussdorff space K, the space $\mathcal{M}(K)$ of all finite complex regular measures on K, the space $\mathcal{M}_c(K)$ of all compactly supported measures in $\mathcal{M}(K)$, the space $\mathcal{M}_b(K)$ of all bounded measures in $\mathcal{M}(K)$, the space $\mathcal{M}_1(K)$ of all probability measures in $\mathcal{M}(K)$, and the space $\mathcal{M}_{1,c}(K)$ of all compactly supported probability measures in $\mathcal{M}(K)$. The point mass concentrated at x is denoted by δ_x. Suppose that we have the following:

(H^*) There is a continuous mapping $(x,y) \mapsto \delta_x * \delta_y$ from $K \times K$ into $\mathcal{M}_{1,c}(K)$. This mapping is called *convolution*.

$(H^\vee)$ There is an involutive homeomorphism $x \mapsto x^\vee$ from K to K. This mapping is called *involution*.

(He) There is a fixed element e in K. This element is called *identity*.

Identifying x by δ_x the mapping in (H^*) has a unique extension to a continuous bilinear mapping from $\mathcal{M}_b(K) \times \mathcal{M}_b(K)$ to $\mathcal{M}_b(K)$. The involution on K extends to a continuous involution on $\mathcal{M}_b(K)$. Convolution maps

91

$\mathcal{M}_1(K) \times \mathcal{M}_1(K)$ into $\mathcal{M}_1(K)$ and involution maps $\mathcal{M}_1(K)$ onto $\mathcal{M}_1(K)$. Then a *DJS hypergroup*, or simply a *hypergroup*, is a quadruple $(K, *, \vee, e)$ satisfying the following axioms: for any x, y, z in K we have

(H1) $\delta_x * (\delta_y * \delta_z) = (\delta_x * \delta_y) * \delta_z$,

(H2) $(\delta_x * \delta_y)^\vee = \delta_{y^\vee} * \delta_{x^\vee}$,

(H3) $\delta_x * \delta_e = \delta_e * \delta_x = \delta_x$,

(H4) e is in the support of $\delta_x * \delta_{y^\vee}$ if and only if $x = y$,

(H5) the mapping $(x, y) \mapsto supp (\delta_x * \delta_y)$ from $K \times K$ into the space of nonvoid compact subsets of K is continuous, the latter being endowed with the Michael topology (see [6]).

For any measures μ, ν in $\mathcal{M}_b(K)$ obviously $\mu * \nu$ denotes their convolution and $\mu^\vee$ denotes the involution of μ. With these operations $\mathcal{M}_b(K)$ is an *algebra with involution*. If $\delta_x * \delta_y = \delta_y * \delta_x$ holds for all x, y in K, then we call the hypergroup *commutative*. If $x^\vee = x$ holds for all x in K, then we call the hypergroup *Hermitian*. By (H2) any Hermitian hypergroup is commutative. For instance, if $K = G$ is a locally compact Haussdorff group, $\delta_x * \delta_y = \delta_{xy}$ for all x, y in K, $x^\vee$ is the inverse of x, and e is the identity of G, then we obviously have a hypergroup $(K, *, \vee, e)$, which is commutative if and only if the group G is commutative. However, not every hypergroup originates in this way.

In any hypergroup K we identify x by δ_x and we define the *right translation operator* τ_y by the element y in K by the formula:

$$\tau_y f(x) = \int_K f \, d(\delta_x * \delta_y),$$

for any f integrable with respect to $\delta_x * \delta_y$. In particular, τ_y is defined for any continuous complex valued function on K. Similarly, we can define *left translation operators*. Sometimes one uses the suggestive notation

$$f(\delta_x * \delta_y) = f(x * y) = \int_K f \, d(\delta_x * \delta_y)$$

for any x, y in K. Obviously, in case of commutative hypergroups the simple term *translation operator* is used. The function $\tau_y f$ is *the translate of f by y*.

The presence of translation operators on commutative hypergroups leads to the concept of variety. Let K be a locally compact Haussdorff space and let $\mathcal{C}(K)$ denote the locally convex topological vector space of all continuous, complex valued functions on K, equipped with the pointwise linear operations and with the topology of uniform convergence on compact sets. The dual of this space can be identified with $\mathcal{M}_c(K)$, the latter being endowed with the weak*-topology with respect to the space of complex valued continuous functions on K (see e.g. [24], Vol. I.). If, in addition, K is equipped with a commutative hypergroup structure, then a subset H of $\mathcal{C}(K)$ is called *translation invariant* if for any f in H the function $\tau_y f$ belongs to H for all y in K. A nonzero, closed translation invariant subspace of $\mathcal{C}(K)$ is called a *variety*. For any f in $\mathcal{C}(K)$ the *variety generated by f* is the closed subspace generated by all translates of f, which is denoted by $\tau(f)$.

If K is a locally compact topological space, then the dual of $\mathcal{C}(K)$ can be identified with the space $\mathcal{M}_c(K)$. It is a locally convex topological vector space. If K is a locally compact topological group, then this space bears a natural algebra structure, corresponding to the convolution of measures (see e.g. [24], Vol. I.). If K is a hypergroup, then it is easy to see (see [6]) that for any continuous function f in $\mathcal{C}(K)$ the function $(x,y) \mapsto f(x*y)$ is continuous. For any measures μ, ν in $\mathcal{M}_c(K)$ and for any f in $\mathcal{C}(K)$,

$$(\mu * \nu)(f) = \int_K \int_K f(x * y) \, d\mu(x) \, d\nu(y)$$

represents the convolution of μ and ν, which is in $\mathcal{M}_c(K)$. The space $\mathcal{M}_c(K)$ equipped with the pointwise linear operations and with the convolution is a commutative algebra with unit.

For any closed linear subspace V in $\mathcal{C}(K)$ its *annihilator* $V^\perp$ in $\mathcal{M}_c(K)$ is the set of all measures from $\mathcal{M}_c(K)$ which vanish on V. Clearly, it is a closed linear subspace of $\mathcal{M}_c(K)$. The dual correspondence is also true: the annihilator $I^\perp$ of any closed linear subspace I of $\mathcal{M}_c(K)$, that is, the set of all elements of $\mathcal{C}(K)$, which belong to the kernel of all linear functionals in I, is a closed linear subspace of $\mathcal{C}(K)$. By the Hahn–Banach Theorem we have the obvious relation $V = V^{\perp\perp}$ for any closed linear subspace V of $\mathcal{C}(K)$. In the case of varieties the annihilators can be characterized (see Section 2.3).

Theorem 7.1. *Let K be a commutative hypergroup, V a variety in $\mathcal{C}(K)$ and I a proper closed ideal in $M(K)$. Then $V^\perp$ is a proper closed ideal in $M(K)$ and $I^\perp$ is a variety in $\mathcal{C}(K)$ (see [66]) .*

Proof. If f is an element of V, μ is an element of $V^\perp$ and ν is arbitrary in $\mathcal{M}_c(K)$, then by definition

$$(\mu * \nu)(f) = \int_K \int_K f(x * y) \, d\mu(x) \, d\nu(y) = \int_K \left[\int_K \tau_y f(x) \, d\mu(x) \right] d\nu(y) = 0,$$

hence $\mu * \nu$ belongs to $V^\perp$. As V is nonzero, $V^\perp$ is a proper ideal. Conversely, if $I \subseteq I^{\perp\perp}$ is a proper closed ideal in $\mathcal{M}_c(K)$, then for any μ in I, f in $I^\perp$ and ν in $\mathcal{M}_c(K)$ we have

$$0 = \int_K f \, d(\mu * \nu) = \int_K \left[\int_K f(x * y) \, d\mu(x) \right] d\nu(y),$$

that is, the function $y \mapsto \mu(\tau_y f)$ annihilates $\mathcal{M}_c(K)$. This means that this function vanishes, and by definition, $\tau_y f$ belongs to $I^\perp$ for all y in K.

Let V be a variety in $\mathcal{C}(K)$. We say that *spectral analysis holds for V* if V contains a minimal (that is, one-dimensional) variety. If spectral analysis holds for any variety in $\mathcal{C}(K)$, then we say that *spectral analysis holds for the hypergroup K*.

Let K be a commutative hypergroup. The function $\varphi : K \to \mathbb{C}$ is called an *exponential* if it is nonidentically zero, and satisfies

$$\varphi(x * y) = \varphi(x)\varphi(y)$$

for all x, y in K. Obviously, for any exponential φ the variety $\tau(\varphi)$ is of one dimension. Conversely, any one-dimensional variety is generated by an exponential. Hence, spectral analysis holds for a given variety V if and only if it contains an exponential. According to Theorem 7.1 spectral analysis holds for a given variety if and only if its annihilator ideal is contained in a maximal ideal generated by an exponential.

Let V be a variety in $\mathcal{C}(K)$. We say that *spectral synthesis holds for V* if V is the sum of finite dimensional varieties. This means that there is a set $(V_\gamma)_{\gamma \in \Gamma}$ of finite dimensional subvarieties of V such that any element f of V can be represented in the form

$$f = f_{\gamma_1} + f_{\gamma_2} + \cdots + f_{\gamma_n}$$

with some positive integer n, with some elements $\gamma_1, \gamma_2, \ldots, \gamma_n$ in Γ and with some functions f_{γ_i} in V_{γ_i}. If K is a locally compact Abelian group, convolution is defined by $\delta_x * \delta_y = \delta_{xy}$, involution is defined by $x^\vee = -x$ and e is the zero element of the group, then finite dimensional varieties are spanned by exponential monomials (see e.g. [62]), hence spectral synthesis holds for a variety in the hypergroup sense if and only if the linear subspace of the variety spanned by its exponential monomials is dense in the variety.

If spectral synthesis holds for any variety in $\mathcal{C}(K)$, then we say that *spectral synthesis holds for the hypergroup K*.

We shall see that — similarly to the case of groups — the Fourier–Laplace transform on $\mathcal{M}_c(K)$ can be used successfully in the study of spectral synthesis. If K is a commutative hypergroup and μ is an element of $\mathcal{M}_c(K)$, then for any exponential φ on K we define

$$\hat{\mu}(\varphi) = \int_K \varphi(x^\vee) \, d\mu(x).$$

Then $\hat{\mu}$ is a complex valued function defined on the set of all exponentials on K. The mapping $\mu \mapsto \hat{\mu}$ is obviously linear. It also has the crucial property

$$(\mu * \nu)\hat{} = \hat{\mu}\,\hat{\nu}$$

for all μ, ν in $\mathcal{M}_c(K)$, which is called the *convolution formula* (see [6]).

An important special class of Hermitian hypergroups is closely related to orthogonal polynomials.

Let $(a_n)_{n \in \mathbb{N}}$, $(b_n)_{n \in \mathbb{N}}$ and $(c_n)_{n \in \mathbb{N}}$ be real sequences with the following properties: $c_n > 0$, $b_n \geq 0$, $a_{n+1} \geq 0$ for all n in $\mathbb{N}$, moreover $a_0 = b_0 = 0$, and $a_n + b_n + c_n = 1$ for all n in $\mathbb{N}$. We define the sequence of polynomials $(P_n)_{n \in \mathbb{N}}$ by $P_0(x) = 1$, $P_1(x) = x$, and by the recursive formula

$$x P_n(x) = a_n P_{n-1}(x) + b_n P_n(x) + c_n P_{n+1}(x)$$

for all $n \geq 1$ and x in $\mathbb{R}$. The following theorem holds (see [6]).

Theorem 7.2. *If the sequence of polynomials $(P_n)_{n \in \mathbb{N}}$ satisfies the above conditions, then there exist constants $c(n, m, k)$ for all n, m, k in $\mathbb{N}$ such that*

$$P_n P_m = \sum_{k=|n-m|}^{n+m} c(n, m, k) P_k$$

holds for all n, m in $\mathbb{N}$.

Proof. By the theorem of J. Favard (see [18], [58]) the conditions on the sequence of polynomials $(P_n)_{n \in \mathbb{N}}$ imply that there exists a probability measure μ on $[-1, 1]$ such that $(P_n)_{n \in \mathbb{N}}$ forms an orthogonal system on $[-1, 1]$ with respect to μ. As P_n has degree n, we have

$$P_n P_m = \sum_{k=0}^{n+m} c(n, m, k) P_k$$

for all n, m in $\mathbb{N}$, where

$$c(n, m, k) = \frac{\int_{-1}^{1} P_k P_n P_m \, d\mu}{\int_{-1}^{1} P_k^2 \, d\mu}$$

holds for all n, m, k in $\mathbb{N}$. The orthogonality of $(P_n)_{n \in \mathbb{N}}$ with respect to μ implies $c(n, m, k) = 0$ for $k > n + m$ or $n > m + k$ or $m > n + k$. Hence our statement is proved.

The formula in the theorem is called *linearization formula*, and the coefficients $c(n, m, k)$ are called *linearization coefficients*. The recursive formula for the sequence $(P_n)_{n \in \mathbb{N}}$ implies $P_n(1) = 1$ for all n in $\mathbb{N}$, hence we have

$$\sum_{k=|n-m|}^{n+m} c(n, m, k) = 1$$

for all n in $\mathbb{N}$. If the linearization is *nonnegative*, that is, the linearization coefficients are nonnegative: $c(n, m, k) \geq 0$ for all n, m, k in $\mathbb{N}$, then we can define a hypergroup structure on $\mathbb{N}$ by the following rule:

$$\delta_n * \delta_m = \sum_{k=|n-m|}^{n+m} c(n, m, k)\delta_k$$

for all n, m in $\mathbb{N}$, with involution as the identity mapping and with e as 0. The resulting discrete Hermitian (hence commutative) hypergroup is called *the polynomial hypergroup associated with the sequence* $(P_n)_{n \in \mathbb{N}}$. We can denote it by $(\mathbb{N}, (P_n)_{n \in \mathbb{N}})$.

We mention here an easy consequence of the linearization formula in polynomial hypergroups. Namely, let $\varphi(n) = P_n^{(k)}(\lambda)$ for all n in $\mathbb{N}$ with some nonnegative integer k and complex number λ. Then we have

$$\varphi(n * 1) = \lambda P_n^{(k)}(\lambda) + k P_n^{(k-1)}(\lambda) = \lambda \varphi(n) + k P_n^{(k-1)}(\lambda)$$

for all n in $\mathbb{N}$. (Here $P_n^{(-1)}$ is meant to be 0.)

7.2 Spectral analysis on polynomial hypergroups in one variable

In this section we show that spectral analysis holds for any polynomial hypergroup in one variable. First we need the general form of exponential functions on polynomial hypergroups, which is well known (see [6], [34]).

Theorem 7.3. *Let K be the polynomial hypergroup associated with the sequence of polynomials $(P_n)_{n \in \mathbb{N}}$. The function $\varphi : \mathbb{N} \to \mathbb{C}$ is an exponential on K if and only if there exists a complex number λ such that*

$$\varphi(n) = P_n(\lambda)$$

holds for all n in $\mathbb{N}$.

Proof. First of all, we remark that if a sequence of polynomials $(P_n)_{n \in \mathbb{N}}$ satisfies a recursion of the form

$$P_n(x)P_m(x) = \sum_{k=0}^{n+m} c(n, m, k) P_k(x)$$

with some real or complex coefficients $c(n, m, k)$ for all real x, then the recursion holds for all complex x. Let λ be a complex number and let $\varphi(n) = P_n(\lambda)$

for any n in $\mathbb{N}$. Then by the definition of convolution we have for any m, n in $\mathbb{N}$,

$$\varphi(\delta_n * \delta_m) = \sum_{k=|n-m|}^{n+m} c(n, m, k)\varphi(k)$$

$$= \sum_{k=|n-m|}^{n+m} c(n, m, k)P_k(\lambda) = P_n(\lambda)P_m(\lambda) = \varphi(n)\varphi(m),$$

hence φ is exponential.

Conversely, let φ be an exponential on K and we define $\lambda = \varphi(1)$. By the exponential property we have for all positive integers n that

$$\lambda\varphi(n) = \varphi(1)\varphi(n) = \varphi(\delta_1 * \delta_n) = \sum_{k=n-1}^{n+1} c(n, 1, k)\varphi(k)$$

$$= c(n, 1, n-1)\varphi(n-1) + c(n, 1, n)\varphi(n) + c(n, 1, n+1)\varphi(n+1).$$

As the same recursion holds for $n \mapsto P_n(\lambda)$, further $\varphi(0) = 1 = P_0(\lambda)$ and $\varphi(1) = \lambda = P_1(\lambda)$, hence $\varphi(n) = P_n(\lambda)$ for all n in $\mathbb{N}$ and the theorem is proved.

Now we can prove that spectral analysis holds for any polynomial hypergroup.

Theorem 7.4. *Spectral analysis holds for any polynomial hypergroup.*

Proof. Let K be the hypergroup associated with the sequence of polynomials $(P_n)_{n \in \mathbb{N}}$ and let V be an arbitrary variety in $\mathcal{C}(K)$. We remark, that $\mathcal{C}(K)$ is the set of all complex valued functions on $\mathbb{N}$, equipped with the pointwise linear operations and with the topology of pointwise convergence. Accordingly, $\mathcal{M}_c(K)$ is the set of all finitely supported complex measures on $\mathbb{N}$. By Theorem 7.1 the annihilator $V^\perp$ of V is a proper closed ideal in $\mathcal{M}_c(K)$. By the convolution formula the Fourier–Laplace transforms of the elements of $V^\perp$ form a proper ideal in the ring of the Fourier–Laplace transforms of all elements of $\mathcal{M}_c(K)$. By Theorem 7.3 the set of all exponentials of K can be identified by $\mathbb{C}$. For any μ in $\mathcal{M}_c(K)$ and for any λ in $\mathbb{C}$ we have

$$\hat{\mu}(\lambda) = \int_{\mathbb{N}} P_n(\lambda)\,d\mu(n).$$

As μ is finitely supported, $\hat{\mu}$ is a complex polynomial on $\mathbb{C}$. We can see easily that any complex polynomial on $\mathbb{C}$ can be written in the form $\hat{\mu}$ with some μ in $\mathcal{M}_c(K)$. Indeed, if p is a complex polynomial on $\mathbb{C}$ of degree n, then it can be written in the form $p = \sum_{k=0}^{n} c_k P_k$ with some complex constants c_k $(k = 0, 1, \ldots, n)$. Then we have

$$p = \Big(\sum_{k=0}^{n} c_k \delta_k \Big)^\widehat{}.$$

This means that the Fourier–Laplace transforms of the elements of $V^\perp$ form a proper ideal in the ring of all complex polynomials on $\mathbb{C}$. By Hilbert's Nullstellensatz (see e.g. [79]) there exists a complex λ_0 such that $\hat{\mu}(\lambda_0) = 0$ for all μ in $V^\perp$, which means by definition that the exponential $n \mapsto P_n(\lambda_0)$ belongs to V and the theorem is proved.

7.3 Spectral synthesis on polynomial hypergroups in one variable

In order to study spectral synthesis on hypergroups, it is necessary to introduce a reasonable concept for exponential monomials. In the group case an exponential monomial can be characterized by the property that the variety it generates contains exactly one exponential and is of finite dimension. One-dimensional varieties are clearly the ones generated by a single exponential. An additive function can be characterized by the properties that it vanishes at zero, the variety it generates is two-dimensional and the only exponential contained in it is the function identically 1. In the case of polynomial hypergroups it seems to be reasonable to introduce the following concept. Let K be the polynomial hypergroup associated with the sequence of polynomials $(P_n)_{n \in \mathbb{N}}$. We call the function $\varphi : \mathbb{N} \to \mathbb{C}$ an *exponential monomial* if it has the form

$$\varphi(n) = \sum_{j=0}^{k} c_j P_n^{(j)}(\lambda)$$

for all n in $\mathbb{N}$, where k is a nonnegative integer and λ, c_j $(j = 0, 1, \ldots, k)$ are complex numbers. The forthcoming theorems will justify this definition. A sum of exponential monomials is called an *exponential polynomial*. It is easy to see (see e.g. [46]) that on a polynomial hypergroup associated with the sequence of polynomials $(P_n)_{n \in \mathbb{N}}$ the additive functions have the form $n \mapsto cP_n'(1)$, where c is any complex constant. Now we list some properties of exponential monomials and polynomials on polynomial hypergroups. In what follows K is a fixed polynomial hypergroup associated with the sequence of polynomials $(P_n)_{n \in \mathbb{N}}$.

Theorem 7.5. *Let k be a nonnegative integer and λ a complex number. Then the functions $n \mapsto P_n^{(j)}(\lambda)$ $(j = 0, 1, \ldots, k)$ are linearly independent.*

Proof. Suppose that

$$c_0 P_n(\lambda) + c_1 P_n'(\lambda) + \cdots + c_k P_n^{(k)}(\lambda) = 0$$

holds for all n in $\mathbb{N}$ with some complex numbers c_j $(j = 0, 1, \ldots, k)$. Substituting $n = 0$ we have $c_0 = 0$. Supposing that we have proved the equations $c_0 = c_1 = \cdots = c_m = 0$ for some $0 \le m < k$, then substituting $n = m + 1$ we have $c_{m+1} = 0$. Hence, by induction we have the statement.

Theorem 7.6. *Let V be a variety over K, k a positive integer, m_i a nonnegative integer and $\lambda_i, c_{i,j}$ complex numbers $(i = 1, 2 \ldots, k, j = 0, 1, \ldots, m_i)$. Suppose that $\lambda_s \ne \lambda_t$ for $s \ne t$. Let*

$$\varphi(n) = \sum_{i=1}^{k} \sum_{j=0}^{m_i} c_{i,j} P_n^{(j)}(\lambda_i)$$

for all n in $\mathbb{N}$. If φ belongs to V, then the function $n \mapsto c_{i,m_i} P_n^{(j)}(\lambda_i)$ belongs to V for $i = 1, 2, \ldots, k$ and $j = 0, 1, \ldots, m_i$.

Proof. Obviously we may suppose that $c_{i,m_i} \ne 0$ for $i = 1, 2, \ldots, k$. We prove the statement by induction on $m_1 + \cdots + m_k$. If $m_1 + \cdots + m_k = 0$, then

$$\varphi(n) = c_{1,0} P_n(\lambda_1) + \cdots + c_{k,0} P_n(\lambda_k)$$

holds for all n in $\mathbb{N}$. We have to show that the function $n \mapsto P_n(\lambda_i)$ belongs to V for $i = 1, 2, \ldots, k$. We prove this statement by induction on k. As it is obvious for $k = 1$ suppose that it has been proved for some $k \ge 1$ and let

$$\varphi(n) = c_{1,0} P_n(\lambda_1) + \cdots + c_{k,0} P_n(\lambda_k) + c_{k+1,0} P_n(\lambda_{k+1})$$

for all n in $\mathbb{N}$.

Let, for all n in $\mathbb{N}$,

$$\psi(n) = \varphi(n * 1) - \lambda_{k+1} \varphi(n).$$

Then ψ belongs to V. On the other hand,

$$\psi(n) = \sum_{i=1}^{k} c_{i,0}(\lambda_i - \lambda_{k+1}) P_n(\lambda_i)$$

holds for all n in $\mathbb{N}$, and hence our statement follows.

Now suppose that the theorem is proved for some $m_1 + \cdots + m_k \ge 1$ and let

$$\varphi(n) = \sum_{i=1}^{k-1} \sum_{j=0}^{m_i} c_{i,j} P_n^{(j)}(\lambda_i) + \sum_{j=0}^{m_k} c_{k,j} P_n^{(j)}(\lambda_k) + c_{k,m_k+1} P_n^{(m_k+1)}(\lambda_k)$$

for all n in $\mathbb{N}$. (For $k = 1$ the first — empty — sum is zero.) Again we set, for all n in $\mathbb{N}$,

$$\psi(n) = \varphi(n * 1) - \lambda_k \varphi(n).$$

Then ψ belongs to V. On the other hand,

$$\psi(n) = \sum_{i=1}^{k-1} \sum_{j=0}^{m_i} c_{i,j}\left[\lambda_i P_n^{(j)}(\lambda_i) + j P_n^{(j-1)}(\lambda_i) - \lambda_k P_n^{(j)}(\lambda_i)\right]$$

$$+ \sum_{j=0}^{m_k} j c_{k,j} P_n^{(j-1)}(\lambda_k) + c_{k,m_k+1}(m_k + 1) P_n^{(m_k)}(\lambda_k)$$

$$= \sum_{i=1}^{k} \sum_{j=0}^{m_k} b_{i,j} P_n^{(j)}(\lambda_i)$$

holds for all n in $\mathbb{N}$, where

$$b_{i,m_i} = c_{i,m_i}(\lambda_i - \lambda_k)$$

for $i = 1, 2, \ldots, k - 1$, and

$$b_{k,m_k} = c_{k,m_k}(m_k + 1),$$

hence our statement follows.

A special case of this theorem is the following.

Theorem 7.7. *Let V be a variety over K, k a positive integer and λ a complex number. If the function $n \mapsto P_n^{(k)}(\lambda)$ belongs to V, then so do the functions $n \mapsto P_n^{(j)}(\lambda)$ for $j = 0, 1, \ldots, k - 1$.*

Another important special case of Theorem 7.6 reads as follows.

Theorem 7.8. *Let K be the polynomial hypergroup associated with the sequence of polynomials $(P_n)_{n \in \mathbb{N}}$. If k is a positive integer, $m_1, m_2, \ldots, m_k$ are nonnegative integers and $\lambda_1, \lambda_2, \ldots, \lambda_k$ are different complex numbers, then the functions $n \mapsto P_n^{(j)}(\lambda_i)$ are linearly independent for $i = 1, 2, \ldots, k$ and $j = 0, 1, \ldots, m_i$.*

Proof. Take $V = \{0\}$ in Theorem 7.6.

Now we show that spectral synthesis holds for any polynomial hypergroup in the sense that the linear hull of all exponential monomials is dense in any variety. Actually, we shall prove that any variety on a polynomial hypergroup is finite dimensional, and it is generated by functions of the form $n \mapsto P_n^{(j)}(\lambda)$ with some finite set of j's and some finite set of λ's. We use the notation of the previous section.

Theorem 7.9. *Spectral synthesis holds for any polynomial hypergroup.*

Proof. Let $(P_n)_{n \in \mathbb{N}}$ be the sequence of polynomials with which the polynomial hypergroup K is associated. First we show that the variety generated by the function $n \mapsto P_n^{(k)}(\lambda)$ is finite dimensional for any nonnegative integer k and for each complex number λ. Let $\psi(n) = P_n^{(k)}(\lambda)$ for any n, k in $\mathbb{N}$ and λ in $\mathbb{C}$. Then by the linearization formula we have

$$\psi(n * m) = \sum_{j=0}^{k} \binom{k}{j} P_n^{(j)}(\lambda) P_m^{(k-j)}(\lambda)$$

for all m, n in $\mathbb{N}$, which yields the statement.

Now we know that for any variety V in $\mathcal{C}(K)$ the Fourier–Laplace transforms of the elements of $V^{\perp}$ form a proper ideal in the ring of all complex polynomials on $\mathbb{C}$. We denote this ideal by J. As J is a principal ideal, it is known that in this case there exist complex numbers $\lambda_1, \lambda_2, \ldots, \lambda_k$ and nonnegative integers $m_1, m_2, \ldots, m_k$ such that a polynomial p belongs to J if and only if $p^{(j)}(\lambda_i) = 0$ holds for $i = 1, 2, \ldots, k$ and $j = 0, 1, \ldots, m_i$. This means that the measure μ in $\mathcal{M}_c(K)$ annihilates V if and only if $\hat{\mu}^{(j)}(\lambda_i) = 0$ holds for $i = 1, 2, \ldots, k$ and $j = 0, 1, \ldots, m_i$, that is, if and only if the functions $n \mapsto P_n^{(j)}(\lambda_i)$ are annihilated by μ for $i = 1, 2, \ldots, k$ and $j = 0, 1, \ldots, m_i$. It follows that V is the closure of the linear hull of these functions. As these functions generate finite dimensional varieties, our statement is proved.

References: [18], [58], [79], [24], [50], [7], [8], [31], [62], [34], [78], [66], [51], [6], [46], [72].

Spectral analysis and synthesis on multivariate polynomial hypergroups

8.1 Polynomial hypergroups in several variables

Let K be a countable set equipped with the discrete topology and let d be a positive integer. We consider a set $(Q_x)_{x \in K}$ of polynomials in d complex variables. If for any nonnegative integer n the symbol K_n denotes the set of all elements x in K for which the degree of Q_x is not greater than n, then we suppose that the polynomials Q_x with x in K_n form a basis for all polynomials of degree not greater than n. In this case for every x, y in K the product $Q_x Q_y$ admits a unique representation

$$Q_x Q_y = \sum_{w \in K} c(x, y, w) Q_w \qquad (8.1)$$

with some complex numbers $c(x, y, w)$. A hypergroup $(K, *)$ is called a *polynomial hypergroup in d variables* or *d-dimensional polynomial hypergroup* if there exists a family of polynomials $(Q_x)_{x \in K}$ in d complex variables satisfying the above condition and such that the convolution in K is defined by

$$\delta_x * \delta_y(\{w\}) = c(x, y, w)$$

for any x, y, w in K. We say that this polynomial hypergroup is *associated with the family of polynomials* $(Q_x)_{x \in K}$. A d-dimensional polynomial hypergroup is also called a *multivariate hypergroup* (see [6]).

It is clear that the polynomial hypergroups in one variable, defined in Section 7.1 represent a special class of this new concept. The above equation (8.1) is a generalization of the linearization formula in Theorem 7.2. It is obvious that any sequence $(p_n)_{n \in \mathbb{N}}$ of polynomials in one variable having the property that for any nonnegative integer n the degree of p_n is exactly n satisfies the above condition in (8.1).

By the conditions on the sequence of polynomials $(Q_x)_{x \in K}$ it follows that there is exactly one element x in K for which Q_x is a nonzero constant. It is easy to see that necessarily $x = e$ is the identity of the hypergroup, and $Q_e = 1$. Sometimes it is convenient to identify the element x in K with the polynomial Q_x. Clearly K contains exactly d nonconstant linear polynomials which are linearly independent. If for some $j = 1, 2, \ldots, d$ there exists an x in K for which $Q_x(z_1, z_2, \ldots, z_d) = z_j$ then we say that the polynomial z_j is in K, and we denote $z_j^\vee = Q_{x^\vee}$.

8.2 Exponential and additive functions on multivariate polynomial hypergroups

In this section first we characterize the exponential functions on the hypergroup K (see [6]).

Theorem 8.1. *Let K be a d-dimensional polynomial hypergroup generated by the family of polynomials $(Q_x)_{x \in K}$. The function $m : K \to \mathbb{C}$ is an exponential if and only if there exists a λ in $\mathbb{C}^d$ such that*

$$m(x) = Q_x(\lambda) \tag{8.2}$$

holds for all x in K.

Proof. The linearization formula and the definition of convolution on K shows immediately that any function m of the form given in (8.2) is an exponential on K.

Conversely, suppose that m is any exponential on K. As the polynomials Q_x for x in K_1 form a basis for all linear polynomials, hence there exists a (unique) λ in $\mathbb{C}^d$ such that (8.2) holds for all x in K_1. We show by induction on the degree of Q_x that this is true for any x in K. Suppose that (8.2) holds for any x in K_n and let x be in K_{n+1}. By our assumptions it follows that Q_x has a representation in the form

$$Q_x(\lambda) = \sum_{j=1}^{s} a_j Q_{x_j}(\lambda) Q_{y_j}(\lambda) \tag{8.3}$$

for any λ in $\mathbb{C}^d$ with some complex numbers a_j and with some x_j in K_1 and y_j in K_n ($j = 1, 2, \ldots, s$), where s is a positive integer. By the definition of the hypergroup structure on K this means that

$$\delta_x = \sum_{j=1}^{s} a_j \delta_{x_j} * \delta_{y_j}$$

holds. Hence we have

$$m(x) = \int_K m \, d\delta_x = \sum_{j=1}^{s} a_j \int_K m \, d(\delta_{x_j} * \delta_{y_j})$$

$$= \sum_{j=1}^{s} a_j m(x_j) m(y_j) = \sum_{j=1}^{s} a_j Q_{x_j}(\lambda) Q_{y_j}(\lambda) = Q_x(\lambda),$$

and our theorem is proved.

This theorem implies that the set of all exponentials of the d-dimensional polynomial hypergroup can be identified with $\mathbb{C}^d$. Consequently, every polynomial hypergroup admits a *normalization* in the sense that there exists a λ_0 in $\mathbb{C}^d$ such that $Q_x(\lambda_0) = 1$ holds for any x in K. Indeed, λ_0 is the unique element in $\mathbb{C}^d$ which corresponds to the exponential 1. We call λ_0 the *normalizing point* of the hypergroup K. In the case of the polynomial hypergroups of one variable we studied in Section 7.1, the normalizing point was 1.

The following theorem describes additive functions on multivariate polynomial hypergroups.

Theorem 8.2. *Let K be a d-dimensional polynomial hypergroup generated by the family of polynomials $(Q_x)_{x \in K}$ with normalizing point λ_0. The function $a : K \to \mathbb{C}$ is additive if and only if there exist complex numbers c_j for $j = 1, 2, \ldots, d$ such that*

$$a(x) = \sum_{i=1}^{d} c_i \partial_i Q_x(\lambda_0) \tag{8.4}$$

holds for all x in K.

Proof. By the linearization formula (8.1),

$$Q_x(\lambda) Q_y(\lambda) = \sum_{w \in K} c(x, y, w) Q_w(\lambda)$$

holds for any x, y in K and for any λ in $\mathbb{C}^d$. Applying ∂_i on both sides of this equation and then substituting $\lambda = \lambda_0$ we have for $i = 1, 2, \ldots, d$,

$$\partial_i Q_x(\lambda_0) + \partial_i Q_y(\lambda_0) = \sum_{w \in K} c(x, y, w) \partial_i Q_w(\lambda_0),$$

which means that the functions $x \mapsto \partial_i Q_x(\lambda_0)$ are additive for $i = 1, 2, \ldots, d$, hence the function a given in (8.4) is additive for any complex numbers $c_1, c_2, \ldots, c_d$.

For the converse, first we observe that the vectors

$$\left(\partial_1 Q_x(\lambda_0), \partial_2 Q_x(\lambda_0), \ldots, \partial_d Q_x(\lambda_0) \right)$$

for x in K_1 and $x \neq e$ are linearly independent because the polynomials Q_x for x in K_1 form a basis for the linear polynomials in d variables. This implies that the system of linear equations

$$a(x) = \sum_{i=1}^{d} c_i \partial_i Q_x(\lambda_0) \tag{8.5}$$

for x in K_1 with $x \neq e$ has a unique solution $c_1, c_2, \ldots, c_d$. Then (8.5) obviously holds also for $x = e$. We show by induction on n that (8.5) holds for any x in K_n and for any n in $\mathbb{N}$. Suppose that this holds for some n and let x be in K_{n+1}. Similarly, as in the proof of the previous theorem, we have that Q_x has a representation in the form (8.3) for all λ in $\mathbb{C}^d$ with some complex numbers a_j and with some x_j in K_1 and y_j in K_n ($j = 1, 2, \ldots, s$), where s is a positive integer, which means that

$$\delta_x = \sum_{j=1}^{s} a_j \delta_{x_j} * \delta_{y_j}$$

holds. On the other hand, applying ∂_i on (8.3) and substituting $\lambda = \lambda_0$ we have for $i = 1, 2, \ldots, d$,

$$\partial_i Q_x(\lambda_0) = \sum_{j=1}^{s} a_j \left(\partial_i Q_{x_j}(\lambda_0) + \partial_i Q_{y_j}(\lambda_0) \right).$$

Finally we obtain

$$a(x) = \int_K a \, d\delta_x = \sum_{j=1}^{s} a_j \int_K a \, d(\delta_{x_j} * \delta_{y_j})$$

$$= \sum_{j=1}^{s} a_j \left(a(x_j) + a(y_j) \right) = \sum_{j=1}^{s} a_j \sum_{i=1}^{d} c_i \left(\partial_i Q_{x_j}(\lambda_0) + \partial_i Q_{y_j}(\lambda_0) \right)$$

$$= \sum_{i=1}^{d} c_i \sum_{j=1}^{s} a_j \left(\partial_i Q_{x_j}(\lambda_0) + \partial_i Q_{y_j}(\lambda_0) \right) = \sum_{i=1}^{d} c_i \partial_i Q_x(\lambda_0),$$

and our theorem is proved.

8.3 Spectral analysis and spectral synthesis on multivariate polynomial hypergroups

In this section we will follow the ideas used in the case of polynomial hypergroups in one variable to prove that spectral analysis and spectral synthesis holds for any multivariate polynomial hypergroup. The basic tool is the characterization of polynomial ideals in $\mathbb{C}[z_1, z_2, \ldots, z_n]$ by differential operators as it is presented in Section 3.4.

First we define exponential monomials and exponential polynomials on multivariate polynomial hypergroups. Using the ideas in the single variable case the following concept seems to be the appropriate one: let K be a multivariate hypergroup in d dimensions associated with the family of polynomials $(Q_x)_{x \in K}$. For any polynomial P in d variables and for any ξ in $\mathbb{C}^d$, the function $x \mapsto P(\partial)Q_x(\xi)$ is called an *exponential monomial* on K and linear combinations of exponential monomials are called *exponential polynomials* on K. By differentiating (8.1) one gets easily that the variety generated by an exponential polynomial is of finite dimension. If V is any variety in $\mathcal{C}(K)$ which contains an exponential, then we say that *spectral analysis holds for V*. If spectral analysis holds for any variety in $\mathcal{C}(K)$, then we say that *spectral analysis holds in K*. If V is any variety in $\mathcal{C}(K)$ and the linear hull of all exponential monomials in V is dense in V, then we say that *spectral synthesis holds for V*. If spectral synthesis holds for any variety in $\mathcal{C}(K)$, then we say that *spectral synthesis holds in K*.

Our observation is based on the Fourier–Laplace transform again. For any finitely supported measure μ in $\mathcal{M}_c(K)$ the modified Fourier–Laplace transform $\widehat{\mu} : \mathbb{C}^d \to \mathbb{C}$ of μ is defined by the equation

$$\widehat{\mu}(\lambda) = \int_K Q_x(\lambda)\, d\mu(x)$$

for any λ in $\mathbb{C}^d$. Then $\widehat{\mu}$ is a polynomial in $\mathbb{C}[z_1, z_2, \ldots, z_n]$. On the other hand, the property of the generating family of polynomials $(Q_x)_{x \in K}$ guarantees that any polynomial in $\mathbb{C}[z_1, z_2, \ldots, z_n]$ is the Fourier–Laplace transform of some finitely supported measure μ in $\mathcal{M}_c(K)$. If V is a variety in $\mathcal{C}(K)$, then its annihilator is a proper ideal in $\mathcal{M}_c(K)$ and the Fourier–Laplace transforms of the measures in this annihilator form a proper ideal in the polynomial ring $\mathbb{C}[z_1, z_2, \ldots, z_n]$. By Hilbert's Nullstellensatz (spectral analysis in $\mathbb{N}^d$) there is a common root ξ of all these Fourier–Laplace transforms, that is

$$\widehat{\mu}(\xi) = \int_K Q_x(\xi)\, d\mu(x) = 0$$

holds for each μ in the annihilator of V, which means that the exponential $x \mapsto Q_x(\xi)$ annihilates the annihilator of V, hence this exponential belongs to V. We have proved the following theorem.

Theorem 8.3. *Spectral analysis holds for any multivariate polynomial hyper-group.*

The corresponding spectral synthesis result follows also easily from Theorem 3.7. Namely, the proper ideal of the Fourier–Laplace transforms of the annihilator of V can be described by conditions given in (3.2), which means that there is a nonempty set Z in $\mathbb{C}^d$ and for any ξ in Z there is a set of polynomials $\mathcal{P}_\xi$ such that the finitely supported measure μ in $\mathcal{M}_c(K)$ annihilates V if and only if

$$P(\partial)\widehat{\mu}(\xi) = \int_K P(\partial)Q_x(\xi)\, d\mu(x) = 0$$

for each ξ in Z and P in $\mathcal{P}_\xi$. This means that the exponential monomials $x \mapsto P(\partial)Q_x(\xi)$ for ξ in Z and P in $\mathcal{M}_c(K)$ belong to V and their linear hull is dense in V. This gives the following result.

Theorem 8.4. *Spectral synthesis holds for any multivariate polynomial hypergroup.*

In particular, any finite dimensional translation invariant function space over a multivariate polynomial hypergroup consists of exponential polynomials. This result is closely related to some results in [17], [41], [42], [43], [62] (see also [66]).

References: [6], [17], [41], [42], [43], [45], [62], [66].

References

1. J. Aczél, *Lectures on functional equations and their applications*, Academic Press, New York, London, 1966.
2. J. J. Benedetto, *Spectral Synthesis*, Academic Press, New York, London, San Fransisco, 1975.
3. A. Bereczky and L. Székelyhidi, *Spectral synthesis on torsion groups*, Jour. Math. Anal. Appl. **304(2)** (2005), 607–613.
4. A. Beurling, *Un théorème sur les fonctions uniformément bornées et continues sur l'axe réel*, Acta Math. **77** (1945), 127–136.
5. A. Beurling, *On the spectral synthesis of bounded functions*, Acta Math. **81** (1949), 225–238.
6. W. R. Bloom and H. Heyer, *Harmonic Analysis of Probability Measures on Hypergroups*, de Gruyter Studies in Mathematics, de Gruyter, Berlin, New York, 1995.
7. A. K. Chilana and K. A. Ross, *Spectral Synthesis in Hypergroups*, Pacific J. Math. **76** (1978), 313–328.
8. A. K. Chilana and A. Kumar, *Spectral synthesis in Segal Algebras on Hypergroups*, Pacific J. Math. **80** (1979), 59–76.
9. S. S. Czerwik, *Functional Equtions and Inequalities in Several Variables*, World Scientific Publishing Co., New Jersey, London, Singapore, Hong Kong, 2002.
10. M. Delsarte, *Les fonctions moyenne-périodiques*, Journal de Math. Pures et Appl. **9(14)** (1935), 403–453.
11. V. Ditkin, *On the structure of ideals in certain normed rings*, Uchenye Zapiski Moskov. Gos. Univ. Matematika **30** (1939), 83–130.
12. D. Z. Djokovič, *A representation theorem for $(X_1 - 1)(X_2 - 1) \ldots (X_n - 1)$ and its applications*, Ann. Polon. Math. **22** (1969), 189–198.
13. L. Ehrenpreis, *Mean periodic functions*, Amer. J. Math. **77** (1955), 293–328.
14. L. Ehrenpreis, *Appendix to "Mean periodic functions"*, Amer. J. Math. **77** (1955), 731–733.
15. R. J. Elliot, *Some results in spectral synthesis*, Proc. Cambridge Phil. Soc. **61** (1965), 214–230.
16. R. J. Elliot, *Two notes on spectral synthesis for discrete Abelian groups*, Proc. Cambridge Phil. Soc. **61** (1965), 617–620.

17. M. Engert, *Finite dimensional translation invariant subspaces*, Pacific J. Math. **32** (1970), 333–343.

18. J. Favard, *Sur les polynomes de Tchebicheff*, C. R. Acad. Sci. Paris **200(A-B)** (1935), 2052–2053.

19. Z. Gajda, (1987), (private communication).

20. J. E. Gilbert, *Two notes on spectral synthesis*, Proc. Cambridge Phil. Soc. **60** (1966), 618–623.

21. J. E. Gilbert, *Spectral synthesis problems for invariant subspaces on groups*, Amer. J. Math. **88** (1966), 626–635.

22. H. Helson, *Spectral synthesis of bounded functions*, Ark. Mat. **1(34)** (1951), 497–502.

23. H. Helson, *Harmonic Analysis*, Addison-Wesley Publishing Company, London, Amsterdam, Sydney, Tokyo, 1983.

24. E. Hewitt and K. Ross, *Abstract Harmonic Analysis I., II.*, Die Grundlehren der Mathematischen Wissenschaften, vol. 115, Springer Verlag, Berlin, Göttingen, Heidelberg, 1963.

25. M. Hosszú, *On a functional equation treated by S. Kurepa*, Glasnik Mat.-Fiz.i Astr. **18** (1963), 59–60.

26. M. Hosszú, *On the functional equation $F(x + y, z) + F(x, y)$ $= F(x, y + z) + F(y, z)$*, Periodica Math. Hungar. **1** (1971), 213–216.

27. D. H. Hyers and Th. M. Rassias, *Approximate homomorphisms*, Aequationes Math. **44** (1992), 125–153.

28. D. H. Hyers and G. Isac and Th. M. Rassias, *Stability of Functional Equations in Several Variables*, Birkhäuser Verlag, Boston, Basel, Berlin, 1998.

29. S. M. Jung, *Hyers–Ulam–Rassias Stability of Functional Equations in Mathematical Analysis*, Hadronic Press Inc., Palm Harbor, Florida, 2001.

30. I. Kaplansky, *Primary ideals in group algebras*, Proc. Nat. Acad. Sci. **35** (1949), 133–136.

31. A. Kumar and A. I. Singh, *Spectral synthesis in products and quotients of hypergroups*, Pacific J. Math. **94** (1981), 179–194.

32. M. Laczkovich and G. Székelyhidi, *Harmonic analysis on discrete Abelian groups*, Proc. Amer. Math. Soc. **133(6)** (2005), 1581–1586.

33. P. G. Laird, *Some properties of mean periodic functions*, Jour. Austral. Math. Soc. **31** (1958), 422–432.

34. R. Lasser, *Orthogonal polynomials and hypergroups*, Rend. Mat. **3** (1983), 185–209.

35. M. Lefranc, *L'analyse harmonique dans $\mathbb{Z}^n$*, C. R. Acad. Sci. Paris **246** (1958), 1951–1953.

36. L. Loomis, *Introduction to Abstract Harmonic Analysis*, Van Nostrand, Princeton, Toronto, London, Melbourne, 1953.

37. J. A. MacDougall and K. Ozeki, *The octahedron equation implies the cube equation: an elementary proof*, Aequationes Math. **31** (1986), 243–246.

38. B. Malgrange, *Sur quelques propriétés des equations des convolutions*, C. R. Acad. Sci, Paris **238** (1954), 2219–2221.

39. P. Malliavin, *Impossibilité de la synthèse spectrale sur les groupes abéliens non compacts*, Inst. des Hautes Etudes Scientifiques, Publications Maths. **2** (1959), 61–68.

40. S. Mandelbrojt and S. Agmon, *Une généralisation du théorème taubérien de Wiener*, Acta Sci. Math. Szeged **12** (1950), 167–176.

41. M. A. McKiernan, *General solution of quadratic functional equations*, Aequationes Math. **3** (1970).

42. M. A. McKiernan, *The matrix equation* $a(x \circ y) = a(x)a(y) + a(y)$, Aequationes Math. **15** (1977), 213–223.

43. M. A. McKiernan, *Equations of the form* $H(x \circ y) = \sum_i f_i(x)g_i(y)$, Aequationes Math. **16** (1977), 51–58.

44. U. Oberst, *The construction of Noetherian operators*, Journal of Algebra **222** (1999), 595–620.

45. Á. Orosz and L. Székelyhidi, *Moment Functions on Polynomial Hypergroups in Several Variables*, Publ. Math. Debrecen **65(3–4)** (2004), 429–438.

46. Á. Orosz and L. Székelyhidi, *Moment Functions on Polynomial Hypergroups*, Arch. Math., **85** (2005), 141–150.

47. *Functional Equations and Inequalities*, (Th. M. Rassias), Kluwer Academic Publishers, Dordrecht, Boston, London, 2000.

48. Th. M. Rassias, *On the stability of functional equations and a problem of Ulam*, Acta Applicandae Mathematicae, **62** (2000), 23–130.

49. J. Riss, *Transformation de Fourier des distributions*, C. R. Acad. Sci. Paris **229** (1949), 12–14.

50. K. A. Ross, *Hypergroups and Centers of Measures Algebras*, Ist. Naz. di alta Mat. **22** (1977), 189–203.

51. K. A. Ross, *Hypergroups and Signed Hypergroups*, Harmonic Analysis on Hypergroups (J. M. Anderson, G. L. Litvinov, K. A. Ross, A. I. Singh, V. S. Sunder, and N. J. Wildberger, eds.), Birkhäuser, Boston, Basel, Berlin, 1998.

52. P. K. Sahoo and L. Székelyhidi, *On a Functional Equation Related to Digital Filtering*, Aequationes Math. **62** (2001), 280–285.

53. P. K. Sahoo and L. Székelyhidi, *A Functional Equation on* $\mathbb{Z}_n \oplus \mathbb{Z}_m$, Acta Math. Hung. **94(1-2)** (2002), 93–98.

54. P. K. Sahoo and L. Székelyhidi, *On the General Solution of a Functional Equation on* $\mathbb{Z} \oplus \mathbb{Z}$, Arch. Math. **81** (2003), 233–239.

55. L. Schwartz, *Théorie génerale des fonctions moyenne-périodiques*, Ann. of Math. **48(2)** (1947), 857–929.

56. L. Schwartz, *Sur une propriété de synthèse spectrale dans les groupes non compact*, C. R. Acad. Sci. Paris **227** (1948), 424–426.

57. I. E. Segal, *The group algebra of a locally compact group*, Trans. Amer. Math. Soc. **61** (1947), 69–105.

58. J. Shohat, *The relation of the classical orthogonal polynomials to the polynomials of Appel*, Amer. J. Math. **58** (1936), 453–464.

59. B. Sturmfels, *Solving Systems of Polynomial Equations*, Lecture Notes in Mathematics, Berkeley, University of California, 2002.

60. L. Sweet, *On the generalized cube and octahedron functional equation*, Aequationes Math. **22** (1981), 29–38.

61. G. Székelyhidi, *Spectral Synthesis on Locally Compact Abelian Groups*, *(essay)*, Cambridge, Trinity College, 2001.

62. L. Székelyhidi, *Note on exponential polynomials*, Pacific J. Math. **103(2)** (1982), 583–587.

63. L. Székelyhidi, *Exponential polynomials and differential equations*, Publ. Math. Debrecen **32** (1985), 105–109.

64. L. Székelyhidi, *The Fourier transform of exponential polynomials*, Publ. Math. Debrecen **33(1-2)** (1986), 13–20.

65. L. Székelyhidi, *The Fourier transform of mean periodic functions*, Util. Math. **29** (1986), 43–48.

66. L. Székelyhidi, *Convolution type functional equations on topological Abelian groups*, World Scientific Publishing Co. Pte. Ltd., Singapore, New Jersey, London, Hong Kong, 1991.

67. L. Székelyhidi, *On common eigenfunctions of difference operators*, Publ. Math. Debrecen **52(3-4)** (1998), 699–704.

68. L. Székelyhidi, *On convolution type functional equations*, Math. Pannonica **10(2)** (1999), 271–275.

69. L. Székelyhidi, *A Wiener Tauberian theorem on discrete Abelian torsion groups*, Annales Acad. Paedag. Cracov., Studia Mathematica I. **4** (2001), 147–150.

70. L. Székelyhidi, *On discrete spectral synthesis*, Functional Equations–Results and Advances (Z. Daróczy and Zs. Páles), Kluwer Academic Publishers, Boston, Dordrecht, London, 2001.

71. L. Székelyhidi, *The Octahedron and Cube Functional Equations Revisited*, Publ. Math. Debrecen **61** (2002), 241–252.

72. L. Székelyhidi, *Spectral Analysis and Spectral Synthesis on Polynomial Hypergroups*, Monatshefte Math. **141(1)** (2004), 33–43.

73. L. Székelyhidi, *The failure of spectral synthesis on some types of discrete Abelian groups*, Jour. Math. Anal. Appl. **291** (2004), 757–763.

74. L. Székelyhidi, *Difference equations via spectral synthesis*, Annales Univ. Sci. Budapest, Sect. Comp., **24** (2004), 3–14.

75. L. Székelyhidi, *Polynomial functions and spectral synthesis*, Aequationes Math., **70(1–2)** (2005), 122–130.

76. L. Székelyhidi, *Spectral synthesis and a characterization of polynomial ideals*, Publ. Math. Debrecen **66(1–2)** (2005), 183–195.

77. L. Székelyhidi, *Spectral Synthesis on Multivariate Polynomial Hypergroups*, (submitted)

78. M. Vogel, *Spectral synthesis on algebras of orthogonal polynomial series*, Math. Z. **194** (1987), 99–116.

79. O. Zariski and P. Samuel, *Commutative Algebra I., II.*, Springer Verlag, New York, Heidelberg, Berlin, 1958, 1960.

Index

0-dimensional, 41

Abelian group, ix–xiv, 8–11, 14, 16, 17, 19, 21–23, 25, 34–37, 49–52, 57–62, 69, 77, 94
absolutely convergent, 3
additive, iv, v, xiii, xiv, 19, 22, 26, 34, 36, 37, 44, 98, 105
additive function, x, 19, 20, 22, 25, 26, 35, 36
adverse, 3
algebra, 4
algebra homomorphism, 1, 2, 4, 9
algebra with involution, 92
algebraic dual, 2
almost everywhere, 9
annihilator, 11, 18, 20, 23, 26, 28, 30, 31, 38, 49, 50, 69, 93, 97, 107, 108
annihilator ideal, 22, 23, 29, 38, 86, 94
antisymmetric, 67
approximate identity, 12, 13
approximation theorem, 10

Baire measure, 10
Banach algebra, 3, 4, 7
Banach space, 3, 4, 12
Banach–Alaoglu Theorem, 4
Beurling problem, 15, 16, 41
bi-additive, 34–37, 44, 61, 67
bijective, 2–4
bilinear, 44, 91
Bochner's Theorem, 9
boundary, 17

canonical representation, 78
character, 9, 14, 18, 26, 38, 61
characteristic differential equation, 87
characteristic differential equation system, 87, 89, 90
characteristic polynomial, 18, 19, 85, 87, 88
characteristic root, 18
closed, ix–xi, 3, 4, 7, 8, 11–15, 17–21, 23, 26, 36, 45, 49, 57, 65, 71, 86
closed ideal, 4, 7, 8, 10–16, 19, 20, 28, 93, 97
closed linear subspace, ix, 72, 77, 93
closed proper ideal, 20
closed subgroup, 50
closed subspace, 10–12, 20, 88, 93
closed translation invariant subspace, 93
closed under differentiation, 32, 33
closure, 4, 10, 15, 101
co-cycle equation, 66, 67
co-final, 40
commutative, 2, 7, 91, 92
commutative algebra, 2, 7, 17, 93
commutative Banach algebra, 3, 4, 7, 8
commutative hypergroup, 91–94, 96
commutative normed algebra, 3
commutative semigroup, 27, 28
commutative topological group, x
compact, ix, x, 4, 7–11, 16, 17, 19, 26, 27, 40, 50, 69, 71, 76, 92
compact Haussdorff space, 4

Springer Monographs in Mathematics

This series publishes advanced monographs giving well-written presentations of the "state-of-the-art" in fields of mathematical research that have acquired the maturity needed for such a treatment. They are sufficiently self-contained to be accessible to more than just the intimate specialists of the subject, and sufficiently comprehensive to remain valuable references for many years. Besides the current state of knowledge in its field, an SMM volume should also describe its relevance to and interaction with neighbouring fields of mathematics, and give pointers to future directions of research.

Jech, T. **Set Theory** (3rd revised edition 2002)

Jorgenson, J.; Lang, S. **Spherical Inversion on SLn (R)** 2001

Kanamori, A. **The Higher Infinite** corr. 2nd printing 2005 (2nd ed. 2003)

Kanovei, V. **Nonstandard Analysis, Axiomatically** 2005

Khoshnevisan, D. **Multiparameter Processes** 2002

Koch, H. **Galois Theory of p-Extensions** 2002

Komornik, V. **Fourier Series in Control Theory** 2005

Kozlov, V.; Maz'ya, V. **Differential Equations with Operator Coefficients** 1999

Lam, T.Y. **Serre's Problem on Projective Modules** 2006

Landsman, N.P. **Mathematical Topics between Classical & QuantumMechanics** 1998

Leach, J.A.; Needham, D.J. **Matched Asymptotic Expansions in Reaction-Diffusion Theory** 2004

Lebedev, L.P.; Vorovich, I.I. **Functional Analysis inMechanics** 2002

Lemmermeyer, F. **Reciprocity Laws: From Euler to Eisenstein** 2000

Malle, G.; Matzat, B.H. **Inverse Galois Theory** 1999

Mardesic, S. **Strong Shape and Homology** 2000

Margulis, G.A. **On Some Aspects of the Theory of Anosov Systems** 2004

Miyake, T. **Modular Forms** 2006

Murdock, J. **Normal Forms and Unfoldings for Local Dynamical Systems** 2002

Narkiewicz, W. **Elementary and Analytic Theory of Algebraic Numbers** 3rd ed. 2004

Narkiewicz, W. **The Development of Prime Number Theory** 2000

Neittaanmaki, P.; Sprekels, J.; Tiba, D. **Optimization of Elliptic Systems. Theory and Applications** 2006

Onishchik, A.L. **Projective and Cayley–Klein Geometries** 2006

Parker, C.; Rowley, P. **Symplectic Amalgams** 2002

Peller, V. (Ed.) **Hankel Operators and Their Applications** 2003

Prestel, A.; Delzell, C.N. **Positive Polynomials** 2001

Puig, L. **Blocks of Finite Groups** 2002

Ranicki, A. **High-dimensional Knot Theory** 1998

Ribenboim, P. **The Theory of Classical Valuations** 1999

Rowe, E.G.P. **Geometrical Physics in Minkowski Spacetime** 2001

Rudyak, Y.B. **On Thorn Spectra, Orientability and Cobordism** 1998

Ryan, R.A. **Introduction to Tensor Products of Banach Spaces** 2002

Saranen, J.; Vainikko, G. **Periodic Integral and Pseudodifferential Equations with Numerical Approximation** 2002

Schneider, P. **Nonarchimedean Functional Analysis** 2002

Serre, J-P. **Complex Semisimple Lie Algebras** 2001 (reprint of first ed. 1987)

Serre, J-P. **Galois Cohomology** corr. 2nd printing 2002 (1st ed. 1997)

Serre, J-P. **Local Algebra** 2000

Serre, J-P. **Trees** corr. 2nd printing 2003 (1st ed. 1980)

Smirnov, E. **Hausdorff Spectra in Functional Analysis** 2002

Springer, T.A.; Veldkamp, F.D. **Octonions, Jordan Algebras, and Exceptional Groups** 2000

Székelyhidi, L. **Discrete Spectral Synthesis and Its Applications** 2006

Sznitman, A.-S. **Brownian Motion, Obstacles and Random Media** 1998

Taira, K. **Semigroups, Boundary Value Problems and Markov Processes** 2003

Talagrand, M. **The Generic Chaining** 2005

Tauvel, P.; Yu, R.W.T. **Lie Algebras and Algebraic Groups** 2005

Tits, J.;Weiss, R.M. **Moufang Polygons** 2002

Uchiyama, A. **Hardy Spaces on the Euclidean Space** 2001

Üstünel, A.-S.; Zakai, M. **Transformation of Measure on Wiener Space** 2000

Vasconcelos, W. **Integral Closure. Rees Algebras, Multiplicities, Algorithms** 2005

Yang, Y. **Solitons in Field Theory and Nonlinear Analysis** 2001

Zieschang, P.-H. **Theory of Association Schemes** 2005